Incognito

Also by David Eagleman

Sum
Why the Net Matters
Wednesday Is Indigo Blue

Incognito

THE SECRET LIVES OF THE BRAIN

David Eagleman

CANONGATE

Edinburgh · London

This paperback edition published by Canongate Books in 2012

3

Copyright © David M. Eagleman, 2011

The moral right of the author has been asserted

First published in Great Britain in 2011 by Canongate Books Ltd,
14 High Street, Edinburgh EH1 1TE

www.canongate.tv

Figure on page 16 © Randy Glasbergen, 2001
Figures on page 25 © Tim Farrell (top) and Ron Rensink (bottom)
Figure on page 29 © Springer
Figure on page 31 © astudio
Figures on page 36 © Fotosearch (left) and Mark Grenier (right)
Figure on page 49 © Elsevier

Every effort has been made to trace copyright holders and obtain their
permission for the use of copyright material. The publisher apologies
for any errors and omissions and would be grateful if notified of
any corrections that should be incorporated into future reprints
or editions of this book.

British Library Cataloguing-in-Publication Data
A catalogue record for this book is available on
request from the British Library

ISBN 978 1 84767 940 6

Typeset in Sabon by Palimpsest Book Production Ltd,
Falkirk, Stirlingshire

Printed and bound in Great Britain by Clays Ltd, St Ives plc

Contents

1. There's Someone In My Head, But It's Not Me   1

2. The Testimony of the Senses: What Is Experience
   Really Like?   20

3. Mind: The Gap   55

4. The Kinds of Thoughts That Are Thinkable   75

5. The Brain Is a Team of Rivals   101

6. Why Blameworthiness Is the Wrong Question   151

7. Life After the Monarchy   193

Appendix   225
Acknowledgments   227
Notes   229
Bibliography   255
Index   281

I

There's Someone In My Head, But It's Not Me

Take a close look at yourself in the mirror. Beneath your dashing good looks churns a hidden universe of networked machinery. The machinery includes a sophisticated scaffolding of interlocking bones, a netting of sinewy muscles, a good deal of specialized fluid, and a collaboration of internal organs chugging away in darkness to keep you alive. A sheet of high-tech self-healing sensory material that we call skin seamlessly covers your machinery in a pleasing package.

And then there's your brain. Three pounds of the most complex material we've discovered in the universe. This is the mission control center that drives the whole operation, gathering dispatches through small portals in the armored bunker of the skull.

Your brain is built of cells called neurons and glia—hundreds of billions of them. Each one of these cells is as complicated as a city. And each one contains the entire human genome and traffics billions of molecules in intricate economies. Each cell sends electrical pulses to other cells, up to hundreds of times per second. If you represented each of these trillions and trillions of pulses in your brain by a single photon of light, the combined output would be blinding.

The cells are connected to one another in a network of such staggering complexity that it bankrupts human language and necessitates new strains of mathematics. A typical neuron makes about ten thousand connections to neighboring neurons. Given the billions of neurons, this means there are as many connections in a single cubic centimeter of brain tissue as there are stars in the Milky Way galaxy.

The three-pound organ in your skull—with its pink consistency of Jell-o—is an alien kind of computational material. It is composed of miniaturized, self-configuring parts, and it vastly outstrips anything we've dreamt of building. So if you ever feel lazy or dull, take heart: you're the busiest, brightest thing on the planet.

Ours is an incredible story. As far as anyone can tell, we're the only system on the planet so complex that we've thrown ourselves headlong into the game of deciphering our own programming language. Imagine that your desktop computer began to control its own peripheral devices, removed its own cover, and pointed its webcam at its own circuitry. That's us.

And what we've discovered by peering into the skull ranks among the most significant intellectual developments of our species: the recognition that the innumerable facets of our behavior, thoughts, and experience are inseparably yoked to a vast, wet, chemical-electrical network called the nervous system. The machinery is utterly alien to us, and yet, somehow, it *is* us.

THE TREMENDOUS MAGIC

In 1949, Arthur Alberts traveled from his home in Yonkers, New York, to villages between the Gold Coast and Timbuktu in West Africa. He brought his wife, a camera, a jeep, and—because of his love of music—a jeep-powered tape recorder. Wanting to open the ears of the western world, he recorded some of the most important music ever to come out of Africa.[1] But Alberts ran into social troubles while using the tape recorder. One West African native heard his voice played back and accused Alberts of "stealing his tongue." Alberts only narrowly averted being pummeled by taking out a mirror and convincing the man that his tongue was still intact.

It's not difficult to see why the natives found the tape recorder so counterintuitive. A vocalization seems ephemeral and ineffable: it is like opening a bag of feathers which scatter on the breeze and can never be retrieved. Voices are weightless and odorless, something you cannot hold in your hand.

So it comes as a surprise that a voice *is* physical. If you build a little machine sensitive enough to detect tiny compressions of the molecules in the air, you can capture these density changes and reproduce them later. We call these machines microphones, and every one of the billions of radios on the planet is proudly serving up bags of feathers once thought irretrievable. When Alberts played the music back from the tape recorder, one West African tribesman depicted the feat as "tremendous magic."

And so it goes with thoughts. What exactly is a thought? It doesn't seem to weigh anything. It feels ephemeral and ineffable. You wouldn't think that a thought has a shape or smell or any sort of physical instantiation. Thoughts seem to be a kind of tremendous magic.

But just like voices, thoughts are underpinned by physical stuff. We know this because alterations to the brain change the kinds of thoughts we can think. In a state of deep sleep, there are no thoughts. When the brain transitions into dream sleep, there are unbidden, bizarre thoughts. During the day we enjoy our normal, well-accepted thoughts, which people enthusiastically modulate by spiking the chemical cocktails of the brain with alcohol, narcotics, cigarettes, coffee, or physical exercise. The state of the physical material determines the state of the thoughts.

And the physical material is absolutely necessary for normal thinking to tick along. If you were to injure your pinkie in an accident you'd be distressed, but your conscious experience would be no different. By contrast, if you were to damage an equivalently sized piece of brain tissue, this might change your capacity to understand music, name animals, see colors, judge risk, make decisions, read signals from your body, or understand the concept of a mirror—thereby unmasking the strange, veiled workings of the machinery beneath. Our hopes, dreams, aspirations, fears, comic instincts, great ideas, fetishes, senses of humor, and desires all emerge from this strange organ—and when the brain changes, so do we. So although it's easy to intuit that thoughts don't have a physical basis, that they are something like feathers on the wind, they in fact depend directly on the integrity of the enigmatic, three-pound mission control center.

The first thing we learn from studying our own circuitry is a simple lesson: most of what we do and think and feel is not under our conscious control. The vast jungles of neurons operate their own programs. The conscious you—the *I* that flickers to life when you wake up in the morning—is the smallest bit of what's transpiring in your brain. Although we are dependent on the functioning of the brain for our inner lives, it runs its own show. Most of its operations are above the security clearance of the conscious mind. The *I* simply has no right of entry.

Your consciousness is like a tiny stowaway on a transatlantic steamship, taking credit for the journey without acknowledging the massive engineering underfoot. This book is about that amazing fact: how we know it, what it means, and what it explains about people, markets, secrets, strippers, retirement accounts, criminals, artists, Ulysses, drunkards, stroke victims, gamblers, athletes, bloodhounds, racists, lovers, and every decision you've ever taken to be yours.

*   *   *

In a recent experiment, men were asked to rank how attractive they found photographs of different women's faces. The photos were eight by ten inches, and showed women facing the camera or turned in three-quarter profile. Unbeknownst to the men, in half the photos the eyes of the women were dilated, and in the other half they were not. The men were consistently more attracted to the women with dilated eyes. Remarkably, the men had no insight into their decision making. None of them said, "I noticed her pupils were two millimeters larger in this photo than in this other one." Instead, they simply felt more drawn toward some women than others, for reasons they couldn't quite put a finger on.

So who was doing the choosing? In the largely inaccessible workings of the brain, *something* knew that a woman's dilated eyes correlates with sexual excitement and readiness. Their brains knew this, but the men in the study didn't—at least not explicitly. The men may also not have known that their notions of beauty and feelings of attraction are deeply hardwired, steered in the right direction by programs carved by millions of years of natural selection. When the men were choosing the most attractive women, they didn't know that the choice was not theirs, *really*, but instead the choice of successful programs that had been burned deep into the brain's circuitry over the course of hundreds of thousands of generations.

Brains are in the business of gathering information and steering behavior appropriately. It doesn't matter whether consciousness is involved in the decision making. And most of the time, it's not. Whether we're talking about dilated eyes, jealousy, attraction, the love of fatty foods, or the great idea you had last week, consciousness is the smallest player in the operations of the brain. Our brains run mostly on autopilot, and the conscious mind has little access to the giant and mysterious factory that runs below it.

You see evidence of this when your foot gets halfway to the brake before you consciously realize that a red Toyota is backing out of a driveway on the road ahead of you. You see it when you notice your name spoken in a conversation across the room that you thought you weren't listening to, when you find someone attractive without knowing why, or when your nervous system gives you a "hunch" about which choice you should make.

The brain is a complex system, but that doesn't mean it's incomprehensible. Our neural circuits were carved by natural selection to solve problems that our ancestors faced during our species' evolutionary history. Your brain has been molded by evolutionary pressures just as your spleen and eyes have been. And so has your consciousness. Consciousness developed because it was advantageous, *but advantageous only in limited amounts.*

Consider the activity that characterizes a nation at any moment. Factories churn, telecommunication lines buzz with activity, businesses ship products. People eat constantly. Sewer lines direct waste. All across the great stretches of land, police chase criminals. Handshakes secure deals. Lovers rendezvous. Secretaries field calls, teachers profess, athletes compete, doctors operate, bus drivers navigate. You may wish to know what's happening at any moment in your great nation, but you can't possibly take in all the information at once. Nor would it be useful, even if you could. You want a summary. So you pick up a newspaper—not a dense paper like the *New York Times* but lighter fare such as *USA Today*. You won't be surprised that none of the details of the activity are listed in the paper; after all, you want to know the bottom line. You want to know that Congress just signed a new tax law that affects your family, but the detailed origin of the idea—involving lawyers and corporations and filibusters—isn't especially important to that new bottom line. And you certainly wouldn't want to know all the details of the food supply of the nation—how the cows are eating and how many are being eaten—you only want to be alerted if there's a spike of mad cow disease. You don't care how the garbage is produced and packed away; you only care if it's going to end up in your backyard. You don't care about the wiring and infrastructure of the factories; you only care if the workers are going on strike. That's what you get from reading the newspaper.

Your conscious mind is that newspaper. Your brain buzzes with activity around the clock, and, just like the nation, almost everything transpires locally: small groups are constantly making decisions and sending out messages to other groups. Out of these local interactions emerge larger coalitions. By the time you read a mental headline, the important action has already transpired, the deals are done. You have surprisingly little access to what happened behind the scenes. Entire political movements gain ground-up support and become unstoppable before you ever catch wind of them as a feeling or an intuition or a thought that strikes you. You're the last one to hear the information.

However, you're an odd kind of newspaper reader, reading the headline and taking credit for the idea as though you thought of it first. You gleefully say, "I just thought of something!", when in fact your brain performed an enormous amount of work before your moment of genius struck. When an idea is served up from behind the scenes, your neural circuitry has been working on it for hours or days or years, consolidating information and trying out new combinations. But you take credit without further wonderment at the vast, hidden machinery behind the scenes.

And who can blame you for thinking you deserve the credit? The brain works its machinations in secret, conjuring ideas like tremendous magic. It does not allow its colossal operating system to be probed by conscious cognition. The brain runs its show incognito.

So who, exactly, deserves the acclaim for a great idea? In 1862, the Scottish mathematician James Clerk Maxwell developed a set of fundamental equations that unified electricity and magnetism. On his deathbed, he coughed up a strange sort of confession, declaring that "something within him" discovered the famous equations, not he. He admitted he had no idea how ideas actually came to him—they simply came to him. William Blake related a similar experience, reporting of his long narrative poem *Milton*: "I have written this poem from immediate dictation twelve or sometimes twenty lines at a time without premeditation and even against my will." Johann Wolfgang von Goethe claimed to have written his novella *The Sorrows of Young Werther* with practically no conscious input, as though he were holding a pen that moved on its own.

And consider the British poet Samuel Taylor Coleridge. He began using opium in 1796, originally for relief from the pain of toothaches and facial neuralgia—but soon he was irreversibly hooked, swigging as much as two quarts of laudanum each week. His poem

"Kubla Khan," with its exotic and dreamy imagery, was written on an opium high that he described as "a kind of a reverie." For him, the opium became a way to tap into his subconscious neural circuits. We credit the beautiful words of "Kubla Khan" to Coleridge because they came from *his* brain and no else's, right? But he couldn't get hold of those words while sober, so who exactly does the credit for the poem belong to?

As Carl Jung put it, "In each of us there is another whom we do not know." As Pink Floyd put it, "There's someone in my head, but it's not me."

*　　*　　*

Almost the entirety of what happens in your mental life is not under your conscious control, and the truth is that it's better this way. Consciousness can take all the credit it wants, but it is best left at the sidelines for most of the decision making that cranks along in your brain. When it meddles in details it doesn't understand, the operation runs less effectively. Once you begin deliberating about where your fingers are jumping on the piano keyboard, you can no longer pull off the piece.

To demonstrate the interference of consciousness as a party trick, hand a friend two dry erase markers—one in each hand—and ask her to sign her name with her right hand at the same time that she's signing it backward (mirror reversed) with her left hand. She will quickly discover that there is only one way she can do it: by *not* thinking about it. By excluding conscious interference, her hands can do the complex mirror movements with no problem—but if she thinks about her actions, the job gets quickly tangled in a bramble of stuttering strokes.

So consciousness is best left uninvited from most of the parties. When it does get included, it's usually the last one to hear the information. Take hitting a baseball. On August 20, 1974, in a game between the California Angels and the Detroit Tigers, the *Guinness Book of World Records* clocked Nolan Ryan's fastball at 100.9 miles per hour (44.7 meters per second). If you work the numbers, you'll see that Ryan's pitch departs the mound and crosses home plate, sixty-feet, six inches away, in four-tenths of a second. This gives just enough time for light signals from the baseball to hit the batter's eye, work through the circuitry of the retina, activate successions of cells along the loopy superhighways of the visual system at the back of the head, cross vast territories to the motor areas, and modify the contraction of the muscles swinging the bat. Amazingly, this entire sequence is possible in less than four-tenths of a second; otherwise no one would ever hit a fastball. But the surprising part is that conscious awareness takes longer than that: about half a second, as we will see in Chapter 2. So the ball travels too rapidly for batters to be consciously aware of it. One does not need to be consciously aware to perform sophisticated motor acts. You can notice this when you begin to duck from a snapping tree branch before you are aware that it's coming toward you, or when you're already jumping up when you first become aware of the phone's ring.

The conscious mind is not at the center of the action in the brain; instead, it is far out on a distant edge, hearing but whispers of the activity.

THE UPSIDE OF DETHRONEMENT

The emerging understanding of the brain profoundly changes our view of ourselves, shifting us from an intuitive sense that we are at the center of the operations to a more sophisticated, illuminating, and wondrous view of the situation. And indeed, we've seen this sort of progress before.

On a starry night in early January 1610, a Tuscan astronomer named Galileo Galilei stayed up late, his eye pressed against the end of a tube he had designed. The tube was a telescope, and it made objects appear twenty times larger. On this night, Galileo observed Jupiter and saw what he thought were three fixed stars near it, strung out on a line across the planet. This formation caught his attention, and he returned to it the following evening. Against his expectations, he saw that all three bodies had moved with Jupiter. That didn't compute: stars don't drift with planets. So Galileo returned his focus to this formation night after night. By January 15 he had cracked the case: these were not fixed stars but, rather, planetary bodies that revolved around Jupiter. Jupiter had moons.

With this observation, the celestial spheres shattered. According to the Ptolemaic theory, there was only a single center—the Earth—around which everything revolved. An alternative idea had been proposed by Copernicus, in which the Earth went around the sun while the moon went around the Earth—but this idea seemed absurd to traditional cosmologists because it required two centers of motion. But here, in this quiet January moment, Jupiter's moons gave testimony to multiple centers: large rocks tumbling in orbit *around* the giant planet could not also be part of the surface of celestial spheres. The Ptolemaic model in which Earth sat at the center of concentric orbits was smashed. The book in which Galileo described his discovery, *Sidereus Nuncius*, rolled off the press in Venice in March 1610 and made Galileo famous.

Six months passed before other stargazers could build instruments with sufficient quality to observe Jupiter's moons. Soon there was a major rush on the telescope-making market, and before long astronomers were spreading around the planet to make a detailed map of our place in the universe. The ensuing four centuries provided an accelerating slide from the center, depositing us firmly as a speck in the visible universe, which contains 500 million galaxy groups, 10 billion large galaxies, 100 billion dwarf galaxies, and 2,000 billion billion suns. (And the visible universe, some 15 billion light-years across, may be a speck in a far larger totality that we cannot yet see.) It is no surprise that these astonishing numbers implied a radically different story about our existence than had been previously suggested.

For many, the fall of the Earth from the center of the universe caused profound unease. No longer could the Earth be considered the paragon of creation: it was now a planet like other planets. This challenge to authority required a change in man's philosophical conception of the universe. Some two hundred years later, Johann Wolfgang von Goethe commemorated the immensity of Galileo's discovery:

> Of all discoveries and opinions, none may have exerted a greater effect on the human spirit. . . . The world had scarcely become known as round and complete in itself when it was asked to waive the tremendous privilege of being the center of the universe. Never, perhaps, was a greater demand made on mankind—for by this admission so many things vanished in mist and smoke! What became of our Eden, our world of innocence, piety and poetry; the testimony of the senses; the conviction of a poetic-religious faith? No wonder his contemporaries did not wish to let all this go and offered every possible resistance to a doctrine which in its converts authorized and demanded a freedom of view and greatness of thought so far unknown, indeed not even dreamed of.

Galileo's critics decried his new theory as a dethronement of man. And following the shattering of the celestial spheres came the shattering of Galileo. In 1633 he was hauled before the Catholic Church's Inquisition, broken of spirit in a dungeon, and forced to scrawl his aggrieved signature on an Earth-centered recantation of his work.[2]

Galileo might have considered himself lucky. Years earlier, another Italian, Giordano Bruno, had also suggested that Earth was not the center, and in February 1600 he was dragged into the public square for his heresies against the Church. His captors, afraid that he might incite the crowd with his famed eloquence, attached an iron mask to his face to prevent him from speaking. He was burned alive at the stake, his eyes peering from behind the mask at a crowd of onlookers who emerged from their homes to gather in the square, wanting to be at the center of things.

Why was Bruno wordlessly exterminated? How did a man with Galileo's genius find himself in shackles on a dungeon floor? Evidently, not everyone appreciates a radical shift of worldview.

If only they could know where it all led! What humankind lost in certainty and egocentrism has been replaced by awe and wonder at our place in the cosmos. Even if life on other planets is terribly unlikely—say the odds are less than one in a billion—we can still expect several billion planets to be sprouting like Chia Pets with life. And if there's only a one-in-a-million chance of life-bearing planets producing meaningful levels of intelligence (say, more than space bacteria), that would still predict several million globes with creatures intermingling in unimaginably strange civilizations. In this way, the fall from the center opened our minds to something much larger.

If you find space science fascinating, strap in for what's happening in brain science: we've been knocked from our perceived position at the center of ourselves, and a much more splendid universe is coming into focus. In this book we'll sail into that inner cosmos to investigate the alien life-forms.

FIRST GLIMPSES INTO THE VASTNESS OF INNER SPACE

Saint Thomas Aquinas (1225–1274) liked to believe that human actions came about from deliberation about what is good. But he couldn't help noticing all the things we do that have little connection with reasoned consideration—such as hiccuping, unconsciously tapping a foot to a rhythm, laughing suddenly at a joke, and so on. This was a bit of a sticking point for his theoretical framework, so he relegated all such actions to a category separate from proper human acts "since they do not proceed from the deliberation of the reason."[3] In defining this extra category, he planted the first seed of the idea of an unconscious.

No one watered this seed for four hundred years, until the polymath Gottfried Wilhelm Leibniz (1646–1716) proposed that the mind is a melding of accessible and inaccessible parts. As a young man, Leibniz composed three hundred Latin hexameters in one morning. He then went on to invent calculus, the binary number system, several new schools of philosophy, political theories, geological hypotheses, the basis of information technology, an equation for kinetic energy, and the first seeds of the idea for software and hardware separation.[4] With all of these ideas pouring out of him, he began to suspect—like Maxwell and Blake and Goethe—that there were perhaps deeper, inaccessible caverns inside him.

Leibniz suggested that there are some perceptions of which we are not aware, and he called these "petite perceptions." Animals have unconscious perceptions, he conjectured—so why can't human beings? Although the logic was speculative, he nonetheless sniffed out that something critical would be left out of the picture if we didn't assume something like an unconscious. "Insensible perceptions are as important to [the science of the human mind] as insensible corpuscles are to natural science," he concluded.[5] Leibniz went on to suggest there were strivings and tendencies ("appetitions") of which we are also unconscious but that can nonetheless drive our actions. This was the first significant exposition of unconscious urges, and he conjectured that his idea would be critical to explaining why humans behave as they do.

He enthusiastically jotted this all down in his *New Essays on Human Understanding*, but the book was not published until 1765, almost half a century after his death. The essays clashed with the Enlightenment notion of knowing oneself, and so they languished unappreciated until almost a century later. The seed sat dormant again.

In the meantime, other events were laying the groundwork for the rise of psychology as an experimental, material science. A Scottish anatomist and theologian named Charles Bell (1774–1842) discovered that nerves—the fine radiations from the spinal cord throughout the body—were not all the same, but instead could be divided into two different kinds: motor and sensory. The former carried information out from the command center of the brain, and the latter brought information back. This was the first major discovery of a pattern to the brain's otherwise mysterious structure, and in the hands of subsequent pioneers this led to a picture of the brain as an organ built with detailed organization instead of shadowy uniformity.

Identifying this sort of logic in an otherwise baffling three-pound block of tissue was highly encouraging, and in 1824 a German philosopher and psychologist named Johann Friedrich Herbart proposed that *ideas themselves* might be understood in a structured mathematical framework: an idea could be opposed by an opposite idea, thus weakening the original idea and causing it to sink below a threshold of awareness.[6] In contrast, ideas that shared a similarity could support each other's rise into awareness. As a new idea climbed, it pulled other similar ones with it. Herbart coined the term "apperceptive mass" to indicate that an idea becomes conscious not in isolation, but only in assimilation with a complex of other ideas already in consciousness. In this way, Herbart introduced a key concept: there exists a *boundary* between conscious and unconscious thoughts; we become aware of some ideas and not of others.

Against this backdrop, a German physician named Ernst Heinrich Weber (1795–1878) grew interested in bringing the rigor of physics to the study of the mind. His new field of "psychophysics" aimed to quantify what people can detect, how fast they can react, and what precisely they perceive.[7] For the first time, perceptions began to be measured with scientific rigor, and surprises began to leak out. For example, it seemed obvious that your senses give you an accurate representation of the outside world—but by 1833 a German physiologist named Johannes Peter Müller (1801–1858) had noticed something puzzling. If he shone light in the eye, put pressure on the eye, or electrically stimulated the nerves of the eye, all of these led to similar sensations of vision—that is, a sensation of *light* rather than of pressure or electricity. This suggested to him that we are not directly aware of the outside world, but instead only of the signals in the nervous system.[8] In other words, when the nervous system tells you that something is "out there"—such as a light—that is what you will believe, irrespective of how the signals get there.

The stage had now been set for people to consider the physical brain as having a relationship with perception. In 1886, years after both Weber and Müller had died, an American named James McKeen Cattell published a paper entitled "The time taken up by cerebral operations."[9] The punch line of his paper was deceptively simple: how quickly you can react to a question depends on the type of thinking you have to do. If you simply have to respond that you've seen a flash or a bang, you can do so quite rapidly (190 milliseconds for flashes and 160 milliseconds for bangs). But if you have to make a choice ("tell me whether you saw a red flash or a green flash"), it takes some tens of milliseconds longer. And if you have to name what you just saw ("I saw a blue flash"), it takes longer still.

Cattell's simple measurements drew the attention of almost no one on the planet, and yet they were the rumblings of a paradigm shift. With the dawning of the industrial age, intellectuals were thinking about *machines*. Just as people apply the computer metaphor now, the machine metaphor permeated popular thought then. By this point, the later part of the nineteenth century, advances in biology had comfortably attributed many aspects of behavior to the machinelike operations of the nervous system. Biologists knew that it took time for signals to be processed in the eyes, travel along the axons connecting them to the thalamus, then ride the nerve highways to the cortex, and finally become part of the pattern of processing throughout the brain.

Thinking, however, continued to be widely considered as

something different. It did not seem to arise from material processes, but instead fell under the special category of the mental (or, often, the spiritual). Cattell's approach confronted the thinking problem head-on. By leaving the stimuli the same but changing the task (*now make such-and-such type of decision*), he could measure how much longer it took for the decision to get made. That is, he could measure *thinking time*, and he proposed this as a straightforward way to establish a correspondence between the brain and the mind. He wrote that this sort of simple experiment brings "the strongest testimony we have to the complete parallelism of physical and mental phenomena; there is scarcely any doubt but that our determinations measure at once the rate of change in the brain and of change in consciousness."[10]

Within the nineteenth-century zeitgeist, the finding that thinking takes time stressed the pillars of the thinking-is-immaterial paradigm. It indicated that thinking, like other aspects of behavior, was not tremendous magic—but instead had a mechanical basis.

Could thinking be equated with the processing done by the nervous system? Could the mind be like a machine? Few people paid meaningful attention to this nascent idea; instead, most continued to intuit that their mental operations appeared immediately at their behest. But for one person, this simple idea changed everything.

ME, MYSELF, AND THE ICEBERG

At the same time that Charles Darwin was publishing his revolutionary book *The Origin of Species*, a three-year-old boy from Moravia was moving with his family to Vienna. This boy, Sigmund Freud, would grow up with a brand-new Darwinian worldview in which man was no different from any other life-form, and the scientific spotlight could be cast on the complex fabric of human behavior.

The young Freud went to medical school, drawn there more by scientific research than clinical application. He specialized in neurology and soon opened a private practice in the treatment of psychological disorders. By carefully examining his patients, Freud came to suspect that the varieties of human behavior were explicable only in terms of unseen mental processes, the machinery running things behind the scenes. Freud noticed that often with these patients there was nothing obvious in their conscious minds driving their behavior, and so, given the new, machinelike view of the brain, he concluded that there must be underlying causes that were hidden from access. In this new view, the mind was not simply equal to the conscious part we familiarly live with; rather it was like an iceberg, the majority of its mass hidden from sight.

This simple idea transformed psychiatry. Previously, aberrant mental processes were inexplicable unless one attributed them to weak will, demon possession, and so on. Freud insisted on seeking the cause in the physical brain. Because Freud lived many decades before modern brain technologies, his best approach was to gather data from the "outside" of the system: by talking to patients and trying to infer their brain states from their mental states. From this vantage, he paid close attention to the information contained in slips of the tongue, mistakes of the pen, behavioral patterns, and the content of dreams. All of these he hypothesized to be the product of hidden neural mechanisms, machinery to which the subject had no direct access. By examining the behaviors poking above the surface, Freud felt confident that he could get a sense of what was lurking below.[11] The more he considered the sparkle from the iceberg's tip, the more he appreciated its depth—and how the hidden mass might explain something about people's thoughts, dreams, and urges.

Applying this concept, Freud's mentor and friend Josef Breuer developed what appeared to be a successful strategy for helping hysterical patients: ask them to talk, without inhibition, about the earliest occurrences of their symptoms.[12] Freud expanded the technique to other neuroses, and suggested that a patient's buried traumatic experiences could be the hidden basis of their phobias, hysterical paralysis, paranoias, and so on. These problems, he guessed, were hidden from the conscious mind. The solution was to draw them up to the level of consciousness so they could be directly confronted and wrung of their neurosis-causing power. This approach served as the basis for psychoanalysis for the next century.

While the popularity and details of psychoanalysis have changed quite a bit, Freud's basic idea provided the first exploration of the way in which hidden states of the brain participate in driving thought and behavior. Freud and Breuer jointly published their work in 1895, but Breuer grew increasingly disenchanted with Freud's emphasis on the sexual origins of unconscious thoughts, and eventually the two parted ways. Freud went on to publish his major exploration of the unconscious, *The Interpretation of Dreams*, in which he analyzed his own emotional crisis and the series of dreams triggered by his father's death. His self-analysis allowed him to reveal unexpected feelings about his father—for example, that his admiration was mixed with hate and shame. This sense of the vast presence below the surface led him to chew on the question of free will. He reasoned that if choices and decisions derive from hidden mental processes, then free choice is either an illusion or, at minimum, more tightly constrained than previously considered.

By the middle of the twentieth century, thinkers began to appreciate that we know ourselves very little. We are not at the center of ourselves, but instead—like the Earth in the Milky Way, and the Milky Way in the universe—far out on a distant edge, hearing little of what is transpiring.

*   *   *

Freud's intuition about the unconscious brain was spot-on, but he lived decades before the modern blossoming of neuroscience. We can now peer into the human cranium at many levels, from electrical spikes in single cells to patterns of activation that traverse the vast territories of the brain. Our modern technology has shaped and focused our picture of the inner cosmos, and in the following chapters we will travel together into its unexpected territories.

How is it possible to get angry at yourself: who, exactly, is mad at whom? Why do rocks appear to climb upward after you stare at a waterfall? Why did Supreme Court Justice William Douglas claim that he was able to play football and go hiking, when everyone could see that he was paralyzed after a stroke? Why was Topsy the elephant electrocuted by Thomas Edison in 1916? Why do people love to store their money in Christmas accounts that earn no interest? If the drunk Mel Gibson is an anti-Semite and the sober Mel Gibson is authentically apologetic, is there a real Mel Gibson? What do Ulysses and the subprime mortgage meltdown have in common? Why do strippers make more money at certain times of month? Why are people whose name begins with J more likely to marry other people whose name begins with J? Why are we so tempted to tell a secret? Are some marriage partners more likely to cheat? Why do patients on Parkinson's medications become compulsive gamblers? Why did Charles Whitman, a high-IQ bank teller and former Eagle Scout, suddenly decide to shoot forty-eight people from the University of Texas Tower in Austin?

What does all this have to do with the behind-the-scenes operations of the brain?

As we are about to see, everything.

2

The Testimony of the Senses:
What Is Experience *Really* Like?

DECONSTRUCTING EXPERIENCE

One afternoon in the late 1800s, the physicist and philosopher Ernst Mach took a careful look at some uniformly colored strips of paper placed next to each other. Being interested in questions of perception, he was given pause by something: the strips did not look quite right. Something was amiss. He separated the strips, looked at them individually, and then put them back together. He finally realized what was going on: although each strip in isolation was uniform in color, when they were placed side by side each appeared to have a gradient of shading: slightly lighter on the left side, and slightly darker on the right. (To prove to yourself that each strip in the figure is in fact uniform in brightness, cover up all but one.)[1]

Mach bands.

Now that you are aware of this illusion of "Mach bands," you'll notice it elsewhere—for example, at the corner where two walls meet, the lighting differences often make it appear that the paint is lighter or darker right next to the corner. Presumably, even though the perceptual fact was in front of you this entire time, you have missed it until now. In the same way, Renaissance painters noticed at some point that distant mountains appeared to be tinted a bit blue—and once this was called out, they began to paint them that way. But the entire history of art up to that point had missed it entirely, even though the data was unhidden in front of them. Why do we fail to perceive these obvious things? Are we really such poor observers of our own experiences?

Yes. We are astoundingly poor observers. And our introspection is useless on these issues: we believe we're seeing the world just fine until it's called to our attention that we're not. We will go through a process of learning to observe our experience, just as Mach carefully observed the shading of the strips. What is our conscious experience *really* like, and what is it not like?

*　　*　　*

Intuition suggests that you open your eyes and voilà: there's the world, with all its beautiful reds and golds, dogs and taxicabs, bustling cities and floriferous landscapes. Vision appears effortless and, with minor exceptions, accurate. There is little important difference, it might seem, between your eyes and a high-resolution digital video camera. For that matter, your ears seem like compact microphones that accurately record the sounds of the world, and your fingertips appear to detect the three-dimensional shape of objects in the outside world. What intuition suggests is dead wrong. So let's see what's really happening.

Consider what happens when you move your arm. Your brain depends on thousands of nerve fibers registering states of contraction and stretching—and yet you perceive no hint of that lightning storm of neural activity. You are simply aware that your limb moved and that it is somewhere else now. Sir Charles Sherrington, an early neuroscience pioneer, spent some time fretting about this fact during the middle of the last century. He was awestruck by the lack of awareness about the vast mechanics under the surface. After all, despite his considerable expertise with nerves, muscles, and tendons, he noted that when he went to pick up a piece of paper, "I have no awareness of the muscles as such at all. . . . I execute the movement rightly and without difficulty."[2] He reasoned that if he were not a neuroscientist it would not have occurred to him to suspect the existence of nerves, muscles, and tendons. This intrigued Sherrington, and he finally inferred that his experience of moving his arm was "a mental product. . . . derived from elements which are not experienced as such and yet . . . the mind uses them in producing the percept." In other words, the storm of nerve and muscle activity is registered by the brain, but what is served up to your awareness is something quite different.

To understand this, let's return to the framework of consciousness as a national newspaper. The job of a headline is to give a tightly compressed summary. In the same manner, consciousness is a way of projecting all the activity in your nervous system into a simpler form. The billions of specialized mechanisms operate below the radar—some collecting sensory data, some sending out motor programs, and the majority doing the main tasks of the neural workforce: combining information, making predictions about what is coming next, making decisions about what to do now. In the face of this complexity, consciousness gives you a summary that is useful for the larger picture, useful at the scale of apples and rivers and humans with whom you might be able to mate.

OPENING YOUR EYES

The act of "seeing" appears so natural that it is difficult to appreciate the vastly sophisticated machinery underlying the process. It may come as a surprise that about one-third of the human brain is devoted to vision. The brain has to perform an enormous amount of work to unambiguously interpret the billions of photons streaming into the eyes. Strictly speaking, all visual scenes are ambiguous: for example, the image to the right can be caused by the Tower of Pisa at a distance of five hundred yards, or a toy model of the tower at arm's length: both cast the identical image on your eyes. Your brain goes through a good deal of trouble to disambiguate the information hitting your eyes by taking context into account, making assumptions,  and using tricks that we'll learn about in a moment. But all this doesn't happen effortlessly, as demonstrated by patients who surgically recover their eyesight after decades of blindness: they do not suddenly see the world, but instead must *learn* to see again.[3] At first the world is a buzzing, jangling barrage of shapes and colors, and even when the optics of their eyes are perfectly functional, their brain must learn how to interpret the data coming in.

For those of us with a lifetime of sight, the best way to appreciate the fact that vision is a construction is by noticing how often our visual systems get it wrong. Visual illusions exist at the edges of what our system has evolved to handle, and as such they serve as a powerful window into the brain.[4]

There is some difficulty in rigorously defining "illusion," as there is a sense in which all of vision is an illusion. The resolution in your peripheral vision is roughly equivalent to looking through a frosted shower door, and yet you enjoy the illusion of seeing the periphery clearly. This is because everywhere you aim your central vision appears to be in sharp focus. To drive this point home, try this demonstration: have a friend hold a handful of colored markers or highlighters out to his side. Keep your gaze fixed on his nose, and now try to name the order of the colors in his hand. The results are surprising: even if you're able to report that there are some colors in your periphery, you won't be able to accurately determine their order. Your peripheral vision is far worse than you would have ever intuited, because under typical circumstances your brain leverages the eye muscles to point your high-resolution central vision directly toward the things you're interested in. Wherever you cast your eyes appears to be in sharp focus, and therefore you assume the whole visual world is in focus.*

That's just the beginning. Consider the fact that we are not aware of the *boundaries* of our visual field. Stare at a point on the wall directly in front of you, stretch your arm out, and wiggle your fingers. Now move your hand slowly back toward your ear. At some point you can no longer see your fingers. Now move it forward again and you can see them. You're crossing the edge of your visual field. Again, because you can always aim your eyes wherever you're interested, you're normally not the least bit aware that there are boundaries beyond which you have no vision. It is interesting to consider that the majority of human beings live their whole lives unaware that they are only seeing a limited cone of vision at any moment.

As we dive further into vision, it becomes clear that your brain can serve up totally convincing perceptions if you simply put the right keys in the right locks. Take the perception of depth. Your two eyes are set a few inches apart, and as a result they receive slightly different images of the world. Demonstrate this to yourself by taking two photographs from a few inches apart, and then putting them side by side. Now cross your eyes so that the two photos merge into a third, and a picture will emerge *in depth*. You will genuinely experience the depth; you can't shake the perception. The impossible notion of depth arising from a flat image divulges the mechanical, automatic nature of the computations in the visual system: feed it the right inputs and it will construct a rich world for you.

One of the most pervasive mistakes is to believe that our visual system gives a faithful representation of what is "out there" in the same way that a movie camera would. Some simple demonstrations

---

*Consider the analogous question of knowing whether your refrigerator light is always on. You might erroneously conclude that it is, simply because it appears that way every time you sneak up to the refrigerator door and yank it open.

Cross your eyes: the two images feed your brain
the illusory signal of depth.

can quickly disabuse you of this notion. In the figure below, two
pictures are shown.

What is the difference between them? Difficult to tell, isn't it?
In a dynamic version of this test, the two images are alternated

Change blindness.

(say, each image shown for half a second, with a tenth of a second
blank period in between). And it turns out we are blind to shockingly large changes in the scene. A large box might be present in

one photo and not the other, or a jeep, or an airplane engine—and the difference goes unseen. Our attention slowly crawls the scene, analyzing interesting landmarks until it finally detects what is changing.* Once the brain has latched onto the appropriate object, the change is easy to see—but this happens only after exhaustive inspection. This "change blindness" highlights the importance of attention: to see an object change, you must attend to it.[5]

You are not seeing the world in the rich detail that you implicitly believed you were; in fact, you are not aware of most of what hits your eyes. Imagine you're watching a short film with a single actor in it. He is cooking an omelet. The camera cuts to a different angle as the actor continues his cooking. Surely you would notice if the actor changed into a different *person*, right? Two-thirds of observers don't.[6]

In one astonishing demonstration of change blindness, random pedestrians in a courtyard were stopped by an experimenter and asked for directions. At some point, as the unsuspecting subject was in the middle of explaining the directions, workmen carrying a door walked rudely right between the two people. Unbeknownst to the subject, the experimenter was stealthily replaced by a confederate who had been hiding behind the door as it was carried: after the door passed, a new person was standing there. The majority of subjects continued giving directions without noticing that the person was not the same as the original one they were talking with.[7] In other words, they were only encoding small amounts of the information hitting their eyes. The rest was assumption.

Neuroscientists weren't the first to discover that placing your eyes on something is no guarantee of seeing it. Magicians figured this out long ago, and perfected ways of leveraging this knowledge.[8] By directing your attention, magicians perform sleight of hand in full view. Their actions *should* give away the game—but they can

The illusion of of "seeing"

*If you haven't spotted it yet, the change in the figure is the height of the wall behind the statue.

rest assured that your brain processes only small bits of the visual scene, not everything that hits your retinas.

This fact helps to explain the colossal number of traffic accidents in which drivers hit pedestrians in plain view, collide with cars directly in front of them, and even intersect unluckily with trains. In many of these cases, the eyes are in the right place, but the brain isn't seeing the stimuli. Vision is more than looking. This also explains why you probably missed the fact that the word "of" is printed twice in the triangle on the previous page.

The lessons here are simple, but they are not obvious, even to brain scientists. For decades, vision researchers barked up the wrong tree by trying to figure out how the visual brain reconstructed a full three-dimensional representation of the outside world. Only slowly did it become clear that the brain doesn't actually use a 3-D model—instead, it builds up something like a 2½-D *sketch* at best.[9] The brain doesn't need a full model of the world because it merely needs to figure out, on the fly, where to look, and when.[10] For example, your brain doesn't need to encode all the details of the coffee shop you're in; it only needs to know how and where to search when it wants something in particular. Your internal model has some general idea that you're in a coffee shop, that there are people to your left, a wall to your right, and that there are several items on the table. When your partner asks, "How many lumps of sugar are left?" your attentional systems interrogate the details of the bowl, assimilating new data into your internal model. Even though the sugar bowl has been in your visual field the entire time, there was no real detail there for your brain. It needed to do extra work to fill in the finer points of the picture.

Similarly, we often know one feature about a stimulus while simultaneously being unable to answer others. Say I were to ask you to look at the following and tell me what it is composed of: ||||||||||||. You would correctly tell me it is composed of vertical lines. If I were to ask you *how many* lines, however, you would be stuck for a while. You can see *that* there are lines, but you cannot tell me *how many* without considerable effort. You can know some things about a scene without knowing other aspects of it, and you become aware of what you're missing only when you're asked the question.

What is the position of your tongue in your mouth? Once you are asked the question you can answer it—but presumably you were not aware of the answer until you asked yourself. The brain generally does not need to know most things; it merely knows how to go out and retrieve the data. It computes on a *need-to-know basis*. You do not continuously track the position of your tongue in consciousness, because that knowledge is useful only in rare circumstances.

In fact, we are not conscious of much of anything until we ask ourselves about it. What does your left shoe feel like on your foot right now? What pitch is the hum of the air conditioner in the background? As we saw with change blindness, we are unaware of most of what should be obvious to our senses; it is only after deploying our attentional resources onto small bits of the scene that we become aware of what we were missing. Before we engage our concentration, we are typically not aware that we are not aware of those details. So not only is our perception of the world a construction that does not accurately represent the outside, but we additionally have the false impression of a full, rich picture when in fact we see only what we need to know, and no more.

The manner in which the brain interrogates the world to gather more details was investigated in 1967 by the Russian psychologist Alfred Yarbus. He measured the exact locations that people were looking at by using an eye tracker, and asked his subjects to gaze at Ilya Repin's painting *An Unexpected Visitor* (next page).[11] The subjects' task was simple: examine the painting. Or, in a different condition, surmise what the people in the painting had been doing just before the "unexpected visitor" came in. Or answer a question about how wealthy the people were. Or their ages. Or how long the unexpected visitor had been away.

The results were remarkable. Depending on what was being asked, the eyes moved in totally different patterns, sampling the picture in a manner that was maximally informative for the question at

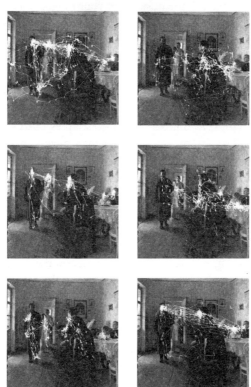

Six records of eye movements from the same subject. Each record lasted three minutes. 1) Free examination. Before subsequent recordings, the subject was asked to: 2) estimate the material circumstances of the family; 3) give the ages of the people; 4) surmise what the family had been doing before the arrival of the "unexpected visitor"; 5) remember the clothes worn by the people; 6) estimate how long the "unexpected visitor" had been away from the family. From Yarbus, 1967.

hand. When asked about the ages of the people, the eyes went to the faces. When asked about their wealth, the focus danced around the clothes and material possessions.

Think about what this means: brains reach out into the world and actively *extract* the type of information they need. The brain does not need to see everything at once about *An Unexpected Visitor*, and it does not need to store everything internally; it only needs to know where to go to find the information. As your eyes interrogate the world, they are like agents on a mission, optimizing their strategy for the data. Even though they are "your" eyes, you have little idea what duty they're on. Like a black ops mission, the eyes operate below the radar, too fast for your clunky consciousness to keep up with.

For a powerful illustration of the limits of introspection, consider the eye movements you are making right now while reading this book. Your eyes are jumping from spot to spot. To appreciate how rapid, deliberate, and precise these eye movements are, just observe someone else while they read. Yet we have no awareness of this active examination of the page. Instead it seems as though ideas simply flow into the head from a stable world.

*   *   *

Because vision appears so effortless, we are like fish challenged to understand water: since the fish has never experienced anything else, it is almost impossible for it to see or conceive of the water. But a bubble rising past the inquisitive fish can offer a critical clue. Like bubbles, visual illusions can call our attention to what we normally take for granted—and in this way they are critical tools for understanding the mechanisms running behind the scenes in the brain.

You've doubtless seen a drawing of a cube like the one to the right. This cube is an example of a "multistable" stimulus—that is, an image that flips back and forth between different perceptions. Pick what you perceive as the "front" face of the cube. Staring at the picture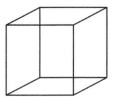

for a moment, you'll notice that sometimes the front face appears to become the back face, and the orientation of the cube changes. If you keep watching, it will switch back again, alternating between these two perceptions of the cube's orientation. There's a striking point here: *nothing has changed on the page, so the change has to be taking place in your brain.* Vision is active, not passive. There is more than one way for the visual system to interpret the stimulus, and so it flips back and forth between the possibilities. The same manner of reversals can be seen in the face–vase illusion below: sometimes you perceive the faces, and sometimes the vase, even though nothing has changed on the page. You simply can't see both at once.

There are even more striking demonstrations of this principle of active vision. Perceptual switching happens if we present one image to your left eye (say, a cow) and a different image to your right eye (say, an airplane). You don't see both at the same time, nor do you see a fusion of the two images—instead, you see one, then the other, then back again.[12] Your visual system is arbitrating a battle between the conflicting information, and you see not what is really out there, but instead only a moment-by-moment version of which perception is winning over the other. Even though the outside world has not changed, your brain dynamically presents different interpretations.

More than actively interpreting what is out there, the brain often goes beyond the call of duty to make things up. Consider the example of the retina, the specialized sheet of photoreceptor cells at the back of the eye. In 1668, the French philosopher and mathematician Edme Mariotte stumbled on something quite unexpected: there is a sizable patch in the retina where the photoreceptors are missing.[13] This missing patch surprised Mariotte because the visual field appears continuous: there is no corresponding gaping hole of vision where the photoreceptors are missing.

Or isn't there? As Mariotte delved more deeply into this issue, he

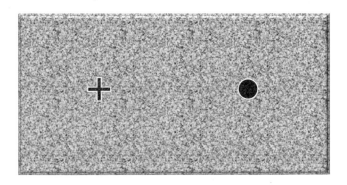

realized that there *is* a hole in our vision—what has come to be known as the "blind spot" in each eye. To demonstrate this to yourself, close your left eye and keep your right eye fixed on the plus sign.

Slowly move the page closer to and farther from your face until the black dot disappears (probably when the page is about twelve inches away). You can no longer see the dot because it is sitting in your blind spot.

Don't assume that your blind spot is small. It's huge. Imagine the diameter of the moon in the night sky. You can fit seventeen moons into your blind spot.

So why hadn't anyone noticed this hole in vision before Mariotte? How could brilliant minds like Michelangelo, Shakespeare, and Galileo have lived and died without ever detecting this basic fact of vision? One reason is because there are two eyes and the blind spots are in different, nonoverlapping locations; this means that with both eyes open you have full coverage of the scene. But more significantly, no one had noticed because the brain "fills in" the missing information from the blind spot. Notice what you see in the location of the dot when it's in your blind spot. When the dot disappears, you do not perceive a hole of whiteness or blackness in its place; instead your brain *invents* a patch of the background pattern. Your brain, with no information from that particular spot in visual space, fills in with the patterns around it.

You're not perceiving what's out there. You're perceiving whatever your brain tells you.

*   *   *

By the mid-1800s, the German physicist and physician Hermann von Helmholtz (1821–1894) had begun to entertain the suspicion that the trickle of data moving from the eyes to the brain is too small to really account for the rich experience of vision. He concluded that the brain must make *assumptions* about the incoming data, and that these assumptions are based on our previous experience.[14] In other words, given a little information, your brain uses its best guesses to turn it into something larger.

Consider this: based on your previous experience, your brain assumes that visual scenes are illuminated by a light source from above.[15] So a flat circle with shading that is lighter at the top and darker at the bottom will be seen as bulging out; one with shading in the opposite direction will be perceived to be dimpling in. Rotating the figure ninety degrees will remove the illusion, making it clear that these are merely flat, shaded circles—but when the figure is turned right side up again, one cannot help but feel an illusory sense of depth.

As a result of the brain's notions about lighting sources, it makes unconscious assumptions about shadows as well: if a square casts a shadow and the shadow suddenly moves, you will believe the square has moved in depth.[16]

Take a look at the figure below: the square hasn't moved at all; the dark square representing its shadow has merely been drawn in a slightly different place. This *could* have happened because the overhead lighting source suddenly shifted position—but because of your previous experience with the slow-moving sun and fixed electrical lighting, your perception automatically gives preference to the likelier explanation: the object has moved toward you.

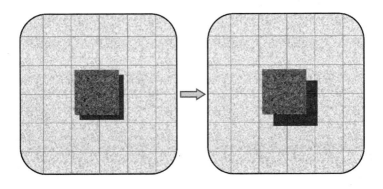

Helmholtz called this concept of vision "unconscious inference," where *inference* refers to the idea that the brain conjectures what might be out there, and *unconscious* reminds us that we have no awareness of the process. We have no access to the rapid and automatic machinery that gathers and estimates the statistics of the world. We're merely the beneficiaries riding on top of the machinery, enjoying the play of light and shadows.

HOW CAN ROCKS DRIFT UPWARD WITHOUT CHANGING POSITION?

When we begin to look closely at that machinery, we find a complex system of specialized cells and circuits in the part of your brain called the visual cortex. There is a division of labor among these circuits: some are specialized for color, some for motion, some for edges, and others for scores of different attributes. These circuits are densely interconnected, and they come to conclusions as a group. When necessary, they serve up a headline for what we might call the *Consciousness Post*. The headline reports only that a bus is coming or that someone has flashed a flirtatious smile—but it does not cite the varied sources. Sometimes it is tempting to think that seeing is easy *despite* the complicated neural machinery that underlies it. To the contrary, it is easy *because of* the complicated neural machinery.

When we take a close look at the machinery, we find that vision can be deconstructed into parts. Stare at a waterfall for a few minutes; after shifting your gaze, stationary objects such as the nearby rocks will briefly appear to crawl upward.[17] Strangely, there is no change in their position over time, even though their movement is clear. Here the imbalanced activity of your motion detectors (usually upward-signaling neurons are balanced in a push–pull relationship with downward-signaling neurons) allows you to see what is impossible in the outside world: motion without position change. This illusion—known as the motion aftereffect or the waterfall illusion—has enjoyed a rich history of study dating back to Aristotle. The illusion illustrates that vision is the product of different modules: in this case, some parts of the visual system insist (incorrectly) that the rocks are moving, while other parts insist that the rocks are not, in fact, changing position. As the philosopher Daniel Dennett has argued, the naïve introspector usually relies on the bad metaphor of the television screen,[18] where moving-while-staying-still cannot happen. But the visual world of the brain is nothing like a television screen, and motion with no change in position is a conclusion it sometimes lands upon.

There are many illusions of motion with no change of position. The figure on the next page demonstrates that static images can appear to move if they happen to tickle motion detectors in the right way. These illusions exist because the exact shading in the pictures stimulates motion detectors in the visual system—and the activity

Motion can be seen even when there is no change in position. (a) High-contrast figures like these stimulate motion detectors, giving the impression of constant movement around the rings. (b) Similarly, the zigzag wheels here appear to turn slowly.

of these receptors is *equivalent* to the perception of motion. If your motion detectors declare that something is moving out there, the conscious you believes it without question. And not merely believes it but *experiences* it.

A striking example of this principle comes from a woman who in 1978 suffered carbon monoxide poisoning.[19] Fortunately, she lived; unfortunately, she suffered irreversible brain damage to parts of her visual system—specifically, the regions involved in representing motion. Because the rest of her visual system was intact, she was able to see stationary objects with no problem. She could tell you there was a ball over there and a telephone over here. But she could no longer see motion. If she stood on a sidewalk trying to cross the street, she could see the red truck over there, and then here a moment later, and finally over there, past her, another moment later—but the truck had no sense of *movement* to it. If she tried to pour water out of a pitcher, she would see a tilted pitcher, then a gleaming column of water hanging from the pitcher, and finally a puddle of water around the glass as it overflowed—but she couldn't see the liquid move. Her life was a series of snapshots. Just as with the waterfall effect, her condition of motion blindness tells us that position and motion are separable in the brain. Motion is "painted on" our views of the world, just as it is erroneously painted on the images above.

A physicist thinks about motion as change in position through time. But the brain has its own logic, and this is why thinking about motion like a physicist rather than like a neuroscientist will lead to wrong predictions about how people operate. Consider baseball outfielders catching fly balls. How do they decide where to run to intercept the ball? Probably their brains represent where the ball is from moment to moment: now it's over there, now it's a little closer, now it's even closer. Right? Wrong.

So perhaps the outfielder's brain calculates the ball's velocity, right? Wrong.

Acceleration? Wrong.

Scientist and baseball fan Mike McBeath set out to understand the hidden neural computations behind catching fly balls.[20] He discovered that outfielders use an unconscious program that tells them not where to end up but simply how to keep running. They move in such a way that the parabolic path of the ball always progresses in a straight line from their point of view. If the ball's path looks like its deviating from a straight line, they modify their running path.

This simple program makes the strange prediction that the outfielders will not dash directly to the landing point of the ball but will instead take a peculiarly curved running path to get there. And that's exactly what players do, as verified by McBeath and his colleagues by aerial video.[21] And because this running strategy gives no information about where the point of intersection will be, only how to keep moving to get there, the program explains why outfielders crash into walls while chasing uncatchable fly balls.

So we see that the system does not need to explicitly represent position, velocity, or acceleration in order for the player to succeed in catching or interception. This is probably not what a physicist would have predicted. And this drives home the point that introspection has little meaningful insight into what is happening behind the scenes. Outfielding greats such as Ryan Braun and Matt Kemp have no idea that they're running these programs; they simply enjoy the consequences and cash the resulting paychecks.

LEARNING TO SEE

When Mike May was three years old, a chemical explosion rendered him completely blind. This did not stop him from becoming the best blind downhill speed skier in the world, as well as a businessman and family man. Then, forty-three years after the explosion robbed him of his vision, he heard about a new surgical development that might be able to restore it. Although he was successful in his life as a blind man, he decided to undergo the surgery.

After the operation, the bandages were removed from around his eyes. Accompanied by a photographer, Mike sat on a chair while his two children were brought in. This was a big moment. It would be the first time he would ever gaze into their faces with his newly cleared eyes. In the resulting photograph, Mike has a pleasant but awkward smile on his face as his children beam at him.

The scene was supposed to be touching, but it wasn't. There was a problem. Mike's eyes were now working perfectly, but he stared with utter puzzlement at the objects in front of him. His brain didn't know what to make of the barrage of inputs. He wasn't experiencing his sons' faces; he was experiencing only uninterpretable sensations of edges and colors and lights. Although his eyes were functioning, he didn't have *vision*.[22]

And this is because the brain has to *learn* how to see. The strange electrical storms inside the pitch-black skull get turned into conscious summaries after a long haul of figuring out how objects in the world match up across the senses. Consider the experience of walking down a hallway. Mike knew from a lifetime of moving down corridors that walls remain parallel, at arm's length, the whole way down. So when his vision was restored, the concept of converging perspective lines was beyond his capacity to understand. It made no sense to his brain.

Similarly, when I was a child I met a blind woman and was amazed at how intimately she knew the layout of her rooms and furniture. I asked her if she would be able to draw out the blueprints with higher accuracy than most sighted people. Her response surprised me: she said she would *not* be able to draw the blueprints at all, because she didn't understand how sighted people converted three dimensions (the room) into two dimensions (a flat piece of paper). The idea simply didn't make sense to her.[23]

Vision does not simply *exist* when a person confronts the world with clear eyes. Instead, an interpretation of the electro-chemical signals streaming along the optic nerves has to be trained up. Mike's brain didn't understand how his own movements changed the sensory consequences. For example, when he moves his head to the left, the scene shifts to the right. The brains of sighted people have come to expect such things and know how to ignore them. But Mike's brain was flummoxed at these strange relationships. And this illustrates a key point: the conscious experience of vision occurs only when there is accurate prediction of sensory consequences,[24] a point to which we will return shortly. So although vision *seems* like a rendition of something that's objectively out there, it doesn't come for free. It has to be learned.

After moving around for several weeks, staring at things, kicking chairs, examining silverware, rubbing his wife's face, Mike came to have the experience of sight as we experience it. He now experiences vision the same way you do. He just appreciates it more.

*   *   *

Mike's story shows that the brain can take a torrent of input and learn to make sense of it. But does this imply the bizarre prediction that you can *substitute* one sense for another? In other words, if you took a data stream from a video camera and converted it into an input to a different sense—taste or touch, say—would you eventually be able to see the world that way? Incredibly, the answer is yes, and the consequences run deep, as we are about to see.

SEEING WITH THE BRAIN

In the 1960s, the neuroscientist Paul Bach-y-Rita at the University of Wisconsin began chewing on the problem of how to give vision to the blind.[25] His father had recently had a miraculous recovery from a stroke, and Paul found himself enchanted by the potential for dynamically reconfiguring the brain.

A question grew in his mind: could the brain substitute one sense for another? Bach-y-Rita decided to try presenting a tactile "display" to blind people.[26] Here's the idea: attach a video camera to someone's forehead and convert the incoming video information into an array of tiny vibrators attached to their back. Imagine putting this device on and walking around a room blindfolded. At first you'd feel a bizarre pattern of vibrations on the small of your back. Although the vibrations would change in strict relation to your own movements, it would be quite difficult to figure out what was going on. As you hit your shin against the coffee table, you'd think, "This really is nothing like vision."

Or isn't it? When blind subjects strap on these visual-tactile substitution glasses and walk around for a week, they become quite good at navigating a new environment. They can translate the feelings on their back into knowing the right way to move. But that's not the stunning part. The stunning part is that they actually begin to perceive the tactile input—to *see* with it. After enough practice, the tactile input becomes more than a cognitive puzzle that needs translation; it becomes a direct sensation.[27]

If it seems strange that nerve signals coming from the back can represent vision, bear in mind that your own sense of vision is carried by nothing but millions of nerve signals that just happen to travel along different cables. Your brain is encased in absolute blackness in the vault of your skull. It doesn't *see* anything. All it knows are these little signals, and nothing else. And yet you perceive the world in all shades of brightness and colors. Your brain is in the dark but your mind constructs light.

To the brain, it doesn't matter where those pulses come from—

from the eyes, the ears, or somewhere else entirely. As long as they consistently correlate with your own movements as you push, thump, and kick things, your brain can construct the direct perception we call vision.[28]

Other sensory substitutions are also under active investigation.[29] Consider Eric Weihenmayer, an extreme rock climber, who scales dangerously sheer rock faces by thrusting his body upward and clinging to precariously shallow foot ledges and handholds. Adding to his feats is the fact that he is blind. He was born with a rare eye disease called retinoschisis, which rendered him blind at thirteen years old. He did not, however, let that crush his dream of being a mountaineer, and in 2001 he became the first (and so far only) blind person to climb Mount Everest. Today he climbs with a grid of over six hundred tiny electrodes in his mouth, called the BrainPort.[30] This device allows him to *see with his tongue* while he climbs. Although the tongue is normally a taste organ, its moisture and chemical environment make it an excellent brain–machine interface when a tingling electrode grid is laid on its surface.[31] The grid translates a video input into patterns of electrical pulses, allowing the tongue to discern qualities usually ascribed to vision, such as distance, shape, direction of movement, and size. The apparatus reminds us that we see not with our eyes but rather with our brains. The technique was originally developed to assist the blind, like Eric, but more recent applications that feed infrared or sonar input to the tongue grid allow divers to see in murky water and soldiers to have 360-degree vision in the dark.[32]

Eric reports that although he first perceived the tongue stimulation as unidentifiable edges and shapes, he quickly learned to recognize the stimulation at a deeper level. He can now pick up a cup of coffee or kick a soccer ball back and forth with his daughter.[33]

If seeing with your tongue sounds strange, think of the experience of a blind person learning to read Braille. At first it's just bumps; eventually those bumps come to have meaning. And if you're having a hard time imagining the transition from cognitive puzzle to direct perception, just consider the way you are reading the letters on this page. Your eyes flick effortlessly over the ornate shapes without any awareness that you are translating them: the meaning of the words simply comes to you. You perceive the language, not the low-level details of the graphemes. To drive home the point, try reading this:

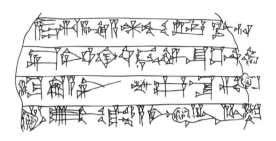

If you were an ancient Sumerian, the meaning would be readily apparent—it would flow off the tablet directly into meaning with no awareness of the mediating shapes. And the meaning of the next sentence is immediately apparent if you're from Jinghong, China (but not from other Chinese regions):

ꨵꨮ ꨵꨰꨣꨳꨵꨮꨮ ꨙꨙꨵꨮ ꨵꨰ ꨵꨯꨵ ꨤꨵꨯꨙꨯꨮ ꨵꨳꨵꨵꨵꨯ ꨵꨵꨮ ꨵꨰ ꨯ.

This next sentence is hilariously funny if you are a reader of the northwestern Iranian language of Baluchi:

توامیں انسان بنی صورتءَ شرپداریں ءُ آجوئیں دروشمءَ ودی بنت ایں۔ اشانی تھا زانت، سرپدی

ءُ شعور است بیت ۔اے وت ما وتا براتی منیل ءُ یکجائیءَ بہ ودیں انت۔

To the reader of cuneiform, New Tai Lue, or Baluchi, the rest of the English script on this page looks as foreign and uninterpretable as their script looks to you. But these letters are effortless for you, because you've already turned the chore of cognitive translation into direct perception.

And so it goes with the electrical signals coming into the brain: at first they are meaningless; with time they accrue meaning. In the same way that you immediately "see" the meaning in these words, your brain "sees" a timed barrage of electrical and chemical signals as, say, a horse galloping between snow-blanketed pine trees. To Mike May's brain, the neural letters coming in are still in need of translation. The visual signals generated by the horse are uninterpretable bursts of activity, giving little indication, if any, of what's out there; the signals on his retina are like letters of Baluchi that struggle to be translated one by one. To Eric Weihenmayer's brain, his tongue is sending messages in New Tai Lue—but with enough practice, his brain learns to understand the language. At that point, his understanding of the visual world is as directly apparent as the words of his native tongue.

Here's an amazing consequence of the brain's plasticity: in the future we may be able to plug new sorts of data streams directly into the brain, such as infrared or ultraviolet vision, or even weather data or stock market data.[34] The brain will struggle to absorb the data at first, but eventually it will learn to speak the language. We'll be able to add new functionality and roll out Brain 2.0.

This idea is not science fiction; the work has already begun. Recently, researchers Gerald Jacobs and Jeremy Nathans took the gene for a human photopigment—a protein in the retina that absorbs light of a particular wavelength—and spliced it into color-blind mice.[35] What emerged? Color vision. These mice can now tell different colors apart. Imagine you give them a task in which they can gain a reward by hitting a blue button but they get no reward for hitting a red button. You randomize the positions of the buttons on each trial. The modified mice, it turns out, learn to choose the blue button, while to normal mice the buttons look indistinguishable—and hence they choose randomly. The brains of the new mice have figured out how to listen to the new dialect their eyes are speaking.

From the natural laboratory of evolution comes a related phenomenon in humans. At least 15 percent of human females possess a genetic mutation that gives them an extra (fourth) type of color photoreceptor—and this allows them to discriminate between colors that look identical to the majority of us with a mere three types of color photoreceptors.[36] Two color swatches that look identical to the majority of people would be clearly distinguishable to these ladies. (No one has yet determined what percentage of fashion arguments is caused by this mutation.)

So plugging new data streams into the brain is not a theoretical notion; it already exists in various guises. It may seem surprising how easily new inputs can become operable—but, as Paul Bach-y-Rita simply summarized his decades of research, "Just give the brain the information and it will figure it out."

If any of this has changed your view of how you perceive reality, strap in, because it gets stranger. We'll next discover why seeing has very little to do with your eyes.

ACTIVITY FROM WITHIN

In the traditionally taught view of perception, data from the sensorium pours into the brain, works its way up the sensory hierarchy, and makes itself seen, heard, smelled, tasted, felt—"perceived." But a closer examination of the data suggests this is incorrect. The brain is properly thought of as a mostly closed system that runs on its own internally generated activity.[37] We already have many examples of this sort of activity: for example, breathing, digestion, and walking are controlled by autonomously running activity generators in your brain stem and spinal cord. During dream sleep the brain is isolated from its normal input, so internal activation is the only source of cortical stimulation. In the awake state, internal activity is the basis for imagination and hallucinations.

The more surprising aspect of this framework is that the internal data is not *generated* by external sensory data but merely *modulated* by it. In 1911, the Scottish mountaineer and neurophysiologist Thomas Graham Brown showed that the program for moving the muscles for walking is built into the machinery of the spinal cord.[38] He severed the sensory nerves from a cat's legs and demonstrated that the cat could walk on a treadmill perfectly well. This indicated that the program for walking was internally generated in the spinal cord and that sensory feedback from the legs was used only to *modulate* the program—when, say, the cat stepped on a slippery surface and needed to stay upright.

The deep secret of the brain is that not only the spinal cord but the entire central nervous system works this way: internally generated activity is modulated by sensory input. In this view, the difference between being awake and being asleep is merely that the data coming in from the eyes *anchors* the perception. Asleep vision (dreaming) is perception that is not tied down to anything in the real world; waking perception is something like dreaming with a little more commitment to what's in front of you. Other examples of unanchored perception are found in prisoners in pitch-dark solitary confinement, or in people in sensory deprivation chambers. Both of these situations quickly lead to hallucinations.

Ten percent of people with eye disease and visual loss will experience visual hallucinations. In the bizarre disorder known as Charles Bonnet syndrome, people losing their sight will begin to see things—such as flowers, birds, other people, buildings—that they know are not real. Bonnet, a Swiss philosopher who lived in the 1700s, first described this phenomenon when he noticed that his grandfather, who was losing his vision to cataracts, tried to interact with objects and animals that were not physically there.

Although the syndrome has been in the literature for centuries, it is underdiagnosed for two reasons. The first is that many physicians do not know about it and attribute its symptoms to dementia. The second is that the people experiencing the hallucinations are discomfited by the knowledge that their visual scene is at least partially the counterfeit coinage of their brains. According to several surveys, most of them will never mention their hallucinations to their doctor out of fear of being diagnosed with mental illness.

As far as the clinicians are concerned, what matters most is whether the patient can perform a reality check and know that he is hallucinating; if so, the vision is labeled a *pseudohallucination*. Of course, sometimes it's quite difficult to know if you're hallucinating. You might hallucinate a silver pen on your desk right now and never suspect it's not real—because its presence is plausible. It's easy to spot a hallucination only when it's bizarre. For all we know, we hallucinate all the time.

As we've seen, what we call normal perception does not really differ from hallucinations, except that the latter are not anchored by external input. Hallucinations are simply unfastened vision.

Collectively, these strange facts give us a surprising way to look at the brain, as we are about to see.

*   *   *

Early ideas of brain function were squarely based on a computer analogy: the brain was an input–output device that moved sensory information through different processing stages until reaching an end point.

But this assembly line model began to draw suspicion when it was discovered that brain wiring does not simply run from A to B to C: there are feedback loops from C to B, C to A, and B to A. Throughout the brain there is as much feedback as feedforward—a feature of brain wiring that is technically called recurrence and colloquially called loopiness.[39] The whole system looks a lot more like a marketplace than an assembly line. To the careful observer, these features of the neurocircuitry immediately raise the possibility that visual perception is not a procession of data crunching that begins from the eyes and ends with some mysterious end point at the back of the brain.

In fact, nested feedback connections are so extensive that the system can even run backward. That is, in contrast to the idea that primary sensory areas merely process input into successively more complex interpretations for the next highest area of the brain, the higher areas are also talking directly back to the lower ones. For instance: shut your eyes and imagine an ant crawling on a red-and-white tablecloth toward a jar of purple jelly. The low-level parts of your visual system just lit up with activity. Even though you weren't actually seeing the ant, you were seeing it in your mind's eye. The higher-level areas were driving the lower ones. So although the eyes feed into these low-level brain areas, the interconnectedness of the system means these areas do just fine on their own in the dark.

It gets stranger. Because of these rich marketplace dynamics, the different senses influence one another, changing the story of what is thought to be out there. What comes in through the eyes is not just the business of the visual system—the rest of the brain is invested as well. In the ventriloquist illusion, sound comes from one location (the ventriloquist's mouth), but your eyes see a moving mouth in a different location (that of the ventriloquist's dummy). Your brain concludes that the sound comes directly from the dummy's mouth. Ventriloquists don't "throw" their voice. Your brain does all of the work for them.

Take the McGurk effect as another example: when the sound of a syllable (*ba*) is synchronized with a video of lip movements mouthing a different syllable (*ga*), it produces the powerful illusion that you are hearing yet a third syllable (*da*). This results from the dense interconnectivity and loopiness in the brain, which allows voice and lip-movement cues to become combined at an early processing stage.[40]

Vision usually dominates over hearing, but a counter example is the illusory flash effect: when a flashed spot is accompanied by two beeps, it appears to flash twice.[41] This is related to another phenomenon called "auditory driving," in which the apparent rate of a flickering light is driven faster or slower by an accompanying beeping sound presented at a different rate.[42] Simple illusions like these serve as powerful clues into neural circuitry, telling us that the visual and auditory systems are densely tied in with each other, trying to relate a unified story of events in the world. The assembly line model of vision in introductory textbooks isn't just misleading, it's dead wrong.

*   *   *

So what is the advantage of a loopy brain? First, it permits an organism to transcend stimulus–response behavior, and instead confers the ability to make predictions ahead of actual sensory input. Think about trying to catch a fly ball. If you were merely an assembly line device, you couldn't do it: there'd be a delay of hundreds of milliseconds from the time light strikes your retina until you could execute a motor command. Your hand would always be reaching for a place where the ball *used* to be. We're able to catch baseballs only because we have deeply hardwired internal models of physics.[43] These internal models generate expectations about when and where the ball will land given the effects of gravitational acceleration.[44] The parameters of the predictive internal models are trained by lifelong exposure in normal, Earth-bound experience. This way, our brains do not work solely from the latest sensory data, but instead construct predictions about where the ball is about to be.

This is a specific example of the broader concept of internal models of the outside world. The brain internally simulates what will happen if you were to perform some action under specific conditions. Internal models not only play a role in motor acts (such as catching or dodging) but also underlie conscious *perception*. As early as the 1940s, thinkers began to toy with the idea that perception works not by building up bits of captured data, but instead by matching *expectations* to incoming sensory data.[45]

As strange as it sounds, this framework was inspired by the observation that our expectations influence what we see. Don't believe it? Try to discern what's in the figure on the following page. If your brain doesn't have a prior expectation about what the blobs mean, you simply see blobs. There has to be a match between your expectations and the incoming data for you to "see" anything.

One of the earliest examples of this framework came from the neuroscientist Donald MacKay, who in 1956 proposed that the visual cortex is fundamentally a machine whose job is to generate a model of the world.[46] He suggested that the primary visual cortex constructs an internal model that allows it to anticipate the data

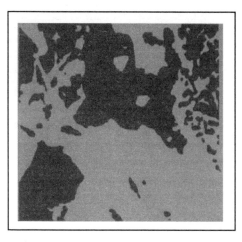

A demonstration of the role of expectation in perception. These blobs generally have no meaning to a viewer initially, and only after a hint does the image make sense. (Don't worry if they still look like blobs to you; a hint comes later in the chapter.) From Ahissar and Hochstein, 2004.

streaming up from the retina (see the appendix for an anatomical guide). The cortex sends its predictions to the thalamus, which reports on the *difference* between what comes in through the eyes and what was already anticipated. The thalamus sends back to the cortex only that difference information—that is, the bit that wasn't predicted away. This unpredicted information adjusts the internal model so there will be less of a mismatch in the future. In this way, the brain refines its model of the world by paying attention to its mistakes. MacKay pointed out that this model is consistent with the anatomical fact that there are ten times as many fibers projecting from the primary visual cortex back to the visual thalamus as there are going the other direction—just what you'd expect if detailed expectations were sent from the cortex to the thalamus and the forward-moving information represented only a small signal carrying the difference.

What all this tells us is that perception reflects the active comparison of sensory inputs with internal predictions. And this gives us a way to understand a bigger concept: awareness of your surroundings occurs only when sensory inputs *violate* expectations. When the world is successfully predicted away, awareness is not needed because the brain is doing its job well. For example, when you first learn how to ride a bicycle, a great deal of conscious concentration is required; after some time, when your sensory-motor predictions have been perfected, riding becomes unconscious. I don't mean you're unaware that you're riding a bicycle, but you *are* unaware of how you're holding the handlebars, applying pressure to the pedals, and balancing your torso. From extensive experience, your brain knows exactly what to expect as you make your movements. So you're conscious neither of the movements nor of the sensations unless something changes—like a strong wind or a flat tire. When these new situations cause your normal expectations to be violated, consciousness comes online and your internal model adjusts.

This predictability that you develop between your own actions and the resulting sensations is the reason you cannot tickle yourself. Other people can tickle you because their tickling maneuvers are not predictable to you. And if you'd really like to, there are ways to take predictability away from your own actions so that you can tickle yourself. Imagine controlling the position of a feather with a time-delay joystick: when you move the stick, at least one second passes before the feather moves accordingly. This takes away the predictability and grants you the ability to self-tickle. Interestingly, schizophrenics can tickle themselves because of a problem with their timing that does not allow their motor actions and resulting sensations to be correctly sequenced.[47]

Recognizing the brain as a loopy system with its own internal dynamics allows us to understand otherwise bizarre disorders. Take Anton's syndrome, a disorder in which a stroke renders a person blind—and the patient *denies* her blindness.[48] A group of doctors will stand around the bedside and say, "Mrs. Johnson, how many of us are around your bed?" and she'll confidently answer, "Four," even though in fact there are seven of them. A doctor will say, "Mrs. Johnson, how many fingers am I holding up?" She'll say,

"Three," while in fact he is holding up none. When he asks, "What color is my shirt?" she'll tell him it is white when it is blue. Those with Anton's syndrome are not *pretending* they are not blind; they truly believe they are not blind. Their verbal reports, while inaccurate, are not lies. Instead, they are experiencing what they take to be vision, but it is all internally generated. Often a patient with Anton's syndrome will not seek medical attention for a little while after the stroke, because she has no idea she is blind. It is only after bumping into enough furniture and walls that she begins to feel that something is amiss. While the patient's answers seem bizarre, they can be understood as her internal model: the external data is not getting to the right places because of the stroke, and so the patient's reality is simply that which is generated by the brain, with little attachment to the real world. In this sense, what she experiences is no different from dreaming, drug trips, or hallucinations.

HOW FAR IN THE PAST DO YOU LIVE?

It is not only vision and hearing that are constructions of the brain. The perception of time is also a construction.

When you snap your fingers, your eyes and ears register information about the snap, which is processed by the rest of the brain. But signals move fairly slowly in the brain, millions of times more slowly than electrons carrying signals in copper wire, so neural processing of the snap takes time. At the moment you perceive it, the snap has already come and gone. Your perceptual world always lags behind the real world. In other words, your perception of the world is like a "live" television show (think *Saturday Night Live*), which is not *actually* live. Instead, these shows are aired with a delay of a few seconds, in case someone uses inappropriate language, hurts himself, or loses a piece of clothing. And so it is with your conscious life: it collects a lot of information before it airs it live.[49]

Stranger still, auditory and visual information are processed at different speeds in the brain; yet the sight of your fingers and the sound of the snap appear simultaneous. Further, your decision to snap *now* and the action itself seem simultaneous with the moment of the snap. Because it's important for animals to get timing right, your brain does quite a bit of fancy editing work to put the signals together in a useful way.

The bottom line is that time is a mental construction, not an accurate barometer of what's happening "out there." Here's a way to prove to yourself that something strange is going on with time: look at your own eyes in a mirror and move your point of focus back and forth so that you're looking at your right eye, then at your left eye, and back again. Your eyes take tens of milliseconds to move from one position to the other, but—here's the mystery—you never see them move. What happens to the gaps in time while your eyes are moving? Why doesn't your brain care about the small absences of visual input?

And the duration of an event—how long it lasted—can be easily distorted as well. You may have noticed this upon glancing at a clock on the wall: the second hand seems to be frozen for slightly too long before it starts ticking along at its normal pace. In the laboratory, simple manipulations reveal the malleability of duration. For example, imagine I flash a square on your computer screen for half a second. If I now flash a second square that is larger, you'll think the second one lasted longer. Same if I flash a square that's brighter. Or moving. These will all be perceived to have a longer duration than the original square.[50]

As another example of the strangeness of time, consider how you know when you performed an action and when you sensed the consequences. If you were an engineer, you would reasonably suppose that something you do at timepoint 1 would result in sensory feedback at timepoint 2. So you would be surprised to discover that in the lab we can make it seem to you as though 2 happens before 1. Imagine that you can trigger a flash of light by pressing a button. Now imagine that we inject a slight delay—say, a tenth of a second—between your press and the consequent flash. After you've pressed the button several times, your brain adapts to this delay, so that the two events seem slightly closer in time. Once you are adapted to the delay, we surprise you by presenting the flash immediately after you press the button. In this condition, you will believe the flash happened before your action: you experience an illusory reversal of action and sensation. The illusion presumably reflects a recalibration of motor-sensory timing which results from a prior expectation that sensory consequences should follow motor acts without delay. The best way to calibrate timing expectations of incoming signals is to interact with the world: each time a person kicks or knocks on something, the brain can make the assumption that the sound, sight, and touch should be simultaneous. If one of the signals arrives with a delay, the brain adjusts its expectations to make it seem as though both events happened closer in time.

Interpreting the timing of motor and sensory signals is not merely a party trick of the brain; it is critical to solving the problem of causality. At bottom, causality requires a temporal order judgment: did my motor act precede or follow the sensory input? The only way this problem can be accurately solved in a multisensory brain is by keeping the expected time of signals well calibrated, so that "before" and "after" can be accurately determined even in the face of different sensory pathways of different speeds.

Time perception is an active area of investigation in my laboratory and others, but the overarching point I want to make here is that our sense of time—how much time passed and what happened when—is constructed by our brains. And this sense is easily manipulated, just like our vision can be.

So the first lesson about trusting your senses is: don't. Just because you *believe* something to be true, just because you *know* it's true, that doesn't mean it *is* true. The most important maxim for fighter pilots is "Trust your instruments." This is because your senses will tell you the most inglorious lies, and if you trust them—instead of your cockpit dials—you'll crash. So the next time someone says, "Who are you going to believe, me or your lying eyes?", consider the question carefully.

After all, we are aware of very little of what is "out there." The brain makes time-saving and resource-saving assumptions and tries to see the world only as well as it needs to. And as we realize that we are not conscious of most things until we ask ourselves questions about them, we have taken the first step in the journey of self-excavation. We see that what we perceive in the outside world is generated by parts of the brain to which we do not have access.

These principles of inaccessible machinery and rich illusion do not apply only to basic perceptions of vision and time. They also apply at higher levels—to what we think and feel and believe—as we shall see in the next chapter.

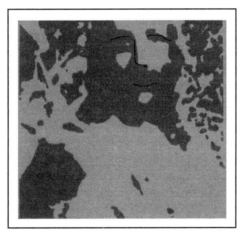

A hint allows the image to take on meaning as a bearded figure. The light patterns hitting your eyes are generally insufficient for vision in the absence of expectations.

3

Mind: The Gap

"I cannot grasp all that I am"

—Augustine

CHANGING LANES

There is a looming chasm between what your brain knows and what your mind is capable of accessing. Consider the simple act of changing lanes while driving a car. Try this: close your eyes, grip an imaginary steering wheel, and go through the motions of a lane change. Imagine that you are driving in the left lane and you would like to move over to the right lane. Before reading on, actually put down the book and try it. I'll give you 100 points if you can do it correctly.

It's a fairly easy task, right? I'm guessing that you held the steering wheel straight, then banked it over to the right for a moment, and then straightened it out again. No problem.

Like almost everyone else, you got it completely wrong.[1] The motion of turning the wheel rightward for a bit, then straightening it out again would steer you off the road: you just piloted a course from the left lane onto the sidewalk. The correct motion for changing lanes is banking the wheel to the right, then back through the center, and continuing to turn the wheel *just as far to the left side*, and only then straightening out. Don't believe it? Verify it for yourself when you're next in the car. It's such a simple motor task that you have no problem accomplishing it in your daily driving. But when forced to access it consciously, you're flummoxed.

The lane-changing example is one of a thousand. You are not consciously aware of the vast majority of your brain's ongoing activities, and nor would you want to be—it would interfere with the brain's well-oiled processes. The best way to mess up your piano piece is to concentrate on your fingers; the best way to get out of breath is to think about your breathing; the best way to miss the golf ball is to analyze your swing. This wisdom is apparent even to children, and we find it immortalized in poems such as "The Puzzled Centipede":

> A centipede was happy quite,
> Until a frog in fun
> Said, "Pray tell which leg comes after which?"
> This raised her mind to such a pitch,
> She lay distracted in the ditch
> Not knowing how to run.

The ability to remember motor acts like changing lanes is called procedural memory, and it is a type of *implicit memory*—meaning that your brain holds knowledge of something that your mind cannot explicitly access.[2] Riding a bike, tying your shoes, typing on a keyboard, or steering your car into a parking space while speaking on your cell phone are examples of this. You execute these actions easily, but without knowing the details of how you do it. You would be totally unable to describe the perfectly timed choreography with which your muscles contract and relax as you navigate around other people in a cafeteria while holding a tray, yet you have no trouble doing it. This is the gap between what your brain can do and what you can tap into consciously.

The concept of implicit memory has a rich, if little known, tradition. By the early 1600s, René Descartes had already begun to suspect that although experience with the world is stored in memory, not all memory is accessible. The concept was rekindled in the late 1800s by the psychologist Hermann Ebbinghaus, who wrote that "most of these experiences remain concealed from consciousness and yet produce an effect which is significant and which authenticates their previous experience."[3]

To the extent that consciousness is useful, it is useful in small quantities, and for very particular kinds of tasks. It's easy to understand why you would not want to be consciously aware of the intricacies of your muscle movement, but this can be less intuitive when applied to your perceptions, thoughts and beliefs, which are also final products of the activity of billions of nerve cells. We turn to these now.

THE MYSTERY OF THE CHICKEN SEXERS AND THE PLANE SPOTTERS

The best chicken sexers in the world hail from Japan. When chicken hatchlings are born, large commercial hatcheries usually set about dividing them into males and females, and the practice of distinguishing the two genders is known as chick sexing. Sexing is necessary because the two genders receive different feeding programs: one for the females, who will eventually produce eggs, and another for the males, who are typically destined to be disposed of because of their uselessness in the commerce of producing eggs; only a few males are kept and fattened for meat. So the job of the chick sexer is to pick up each hatchling and quickly determine its sex in order to choose the correct bin to put it in. The problem is that the task is famously difficult: male and female chicks look exactly alike.

Well, almost exactly. The Japanese invented a method of sexing chicks known as vent sexing, by which expert chicken sexers could rapidly ascertain the sex of one-day-old hatchlings. Beginning in the 1930s, poultry breeders from around the world traveled to the Zen-Nippon Chick Sexing School in Japan to learn the technique.

The mystery was that no one could explain exactly how it was done.[4] It was somehow based on very subtle visual cues, but the professional sexers could not report what those cues were. Instead, they would look at the chick's rear (where the vent is) and simply seem to *know* the correct bin to throw it in.

And this is how the professionals taught the student sexers. The master would stand over the apprentice and watch. The students would pick up a chick, examine its rear, and toss it into one bin or the other. The master would give feedback: *yes* or *no*. After weeks on end of this activity, the student's brain was trained up to masterful—albeit unconscious—levels.

Meanwhile, a similar story was unfolding oceans away. During World War II, under constant threat of bombings, the British had a great need to distinguish incoming aircraft quickly and accurately. Which aircraft were British planes coming home and which were German planes coming to bomb? Several airplane enthusiasts had proved to be excellent "spotters," so the military eagerly employed their services. These spotters were so valuable that the government quickly tried to enlist more spotters—but they turned out to be rare and difficult to find. The government therefore tasked the spotters with training others. It was a grim attempt. The spotters tried to explain their strategies but failed. No one got it, not even the spotters themselves. Like the chicken sexers, the spotters had little idea how they did what they did—they simply saw the right answer.

With a little ingenuity, the British finally figured out how to successfully train new spotters: by trial-and-error feedback. A novice would hazard a guess and the expert would say *yes* or *no*. Eventually the novices became, like their mentors, vessels of the mysterious, ineffable expertise.[5]

There can be a large gap between knowledge and awareness. When we examine skills that are not amenable to introspection, the first surprise is that implicit memory is completely separable from explicit memory: you can damage one without hurting the other. Consider patients with anterograde amnesia, who cannot consciously recall new experiences in their lives. If you spend an afternoon trying to teach them the video game Tetris, they will tell you the next day that they have no recollection of the experience, that they have never seen this video game before, and, most likely, that they have no idea who you are, either. But if you look at their *performance* on the game the next day, you'll find that they have improved exactly as much as nonamnesiacs.[6] Implicitly their brains have learned the game—the knowledge is simply not accessible to their consciousness. (Interestingly, if you wake up an amnesic patient during the night after they've played Tetris, they'll report that they were dreaming of colorful falling blocks, but they have no idea why.)

Of course, it's not just sexers and spotters and amnesiacs who enjoy unconscious learning: essentially everything about your interaction with the world rests on this process.[7] You may have a difficult time putting into words the characteristics of your father's walk, or the shape of his nose, or the way he laughs—but when you see someone who walks, looks, or laughs like him, you know it immediately.

HOW TO KNOW IF YOU'RE A RACIST

We often do not know what's buried in the caverns of our unconscious. An example of this comes up, in its ugliest form, with racism.

Consider this situation: A white company owner refuses employment to a black applicant, and the case goes to court. The employer insists that he harbors no racism; the applicant insists otherwise. The judge is stuck: how can one ever know what sort of biases may lurk in someone's unconscious, modulating their decisions, even if they are not aware of it consciously? People don't always speak their minds, in part because people don't always *know* their minds. As E. M. Forster quipped: "How do I know what I think until I hear what I say?"

But if someone is unwilling to *say* something, are there ways of probing what is in the unconscious brain? Are there ways to ferret out subterranean beliefs by observing someone's behavior?

Imagine that you sit down in front of two buttons, and you're asked to hit the right button whenever a positive word flashes on the screen (*joy*, *love*, *happy*, and so on), and the left button whenever you see a negative word (*terrible*, *nasty*, *failure*). Pretty straightforward. Now the task changes a bit: hit the right button whenever you see a photo of an overweight person, and the left button whenever you see a photo of a thin person. Again, pretty easy. But for the next task, things are paired up: you're asked to hit the right button when you see either a positive word *or* an overweight person, and the left button whenever you see a negative word *or* a thin person. In another group of trials, you do the same thing but with the pairings switched—so you now press the right button for a negative word *or* a thin person.

The results can be troubling. The reaction times of subjects are faster when the pairings have a strong association unconsciously.[8] For example, if overweight people are linked with a negative association in the subject's unconscious, then the subject reacts faster to a photo of an overweight person when the response is linked to the same button as a negative word. During trials in which the opposite concepts are linked (thin with bad), subjects will take a longer time to respond, presumably because the pairing is more difficult. This experiment has been modified to measure implicit attitudes toward races, religions, homosexuality, skin tone, age, disabilities, and presidential candidates.[9]

Another method for teasing out implicit biases simply measures the way a participant moves a computer cursor.[10] Imagine that you start with your cursor positioned at the bottom of the screen, and in the upper corners of the screen you have buttons labeled "like" and "dislike". Then a word appears in the middle (say, the name of a religion), and you are instructed to move the mouse as quickly as you can to your answer about whether you like or dislike people of that creed. What you don't realize is that the exact *trajectory* of your mouse movement is being recorded—every position at every moment. By analyzing the path your mouse traveled, researchers can detect whether your motor system started moving toward one button before other cognitive systems kicked into gear and drove it toward the other response. So, for example, even if you answered "like" for a particular religion, it may be that your trajectory drifted slightly toward the "dislike" button before it got back on track for the more socially appropriate response.

Even people with certainty about their attitudes toward different races, genders, and religions can find themselves surprised—and appalled—by what's lurking in their brains. And like other forms of implicit association, these biases are impenetrable to conscious introspection.*

HOW DO I LOVE THEE?
LET ME COUNT THE *J*'S

Let's consider what happens when two people fall in love. Common sense tells us that their ardor grows from any number of seeds, including life circumstances, a sense of understanding, sexual attraction, and mutual admiration. Surely the covert machinery of the unconscious is not implicated in who you choose as a mate. Or isn't it?

Imagine you run into your friend Joel, and he tells you that he has found the love of his life, a woman named Jenny. That's funny, you consider, because your friend Alex just married Amy, and Donny is crazy for Daisy. Is there something going on with

*It is currently an open question whether courts of law will allow these tests to be admitted as evidence—for example, to probe whether an employer (or attacker or murderer) shows signs of racism. At the moment it is probably best if these tests remain outside the courtroom, for while complicated human decisions are biased by inaccessible associations, it is difficult to know how much these biases influence our final behavior. For example, someone may override their racist biases by more socialized decision-making mechanisms. It is also the case that someone may be a virulent racist, but that was not their reason for a particular crime.

these letter pairings? Is like attracted to like? That's crazy, you conclude: important life decisions—such as who to spend your life with—can't be influenced by something as capricious as the first letter of a name. Perhaps all these alliterative alliances are just an accident.

But they're not an accident. In 2004, psychologist John Jones and his colleagues examined fifteen thousand public marriage records from Walker County, Georgia, and Liberty County, Florida. They found that, indeed, people more often get married to others with the same first letter of their first name than would be expected by chance.[11]

But why? It's not about the letters, exactly—instead it's about the fact that those mates somehow remind their spouses of themselves. People tend to love reflections of themselves in others. Psychologists interpret this as an unconscious self-love, or perhaps a comfort level with things that are familiar —and they term this *implicit egotism.*

Implicit egotism is not just about life partners—it also influences the products you prefer and purchase. In one study, subjects were presented with two (fictional) brands of tea to taste-test. One of the brand names of the teas happened to share its first three letters with the subject's name; that is, Tommy might be sampling teas named Tomeva and Lauler. Subjects would taste the teas, smack their lips, consider both carefully, and almost always decide that they preferred the tea whose name happened to match the first letters of their name. Not surprisingly, a subject named Laura would choose the tea named Lauler. They weren't explicitly *aware* of the connection with the letters; they simply believed the tea tasted better. As it turns out, both cups of tea had been poured from the same teapot.

The power of implicit egotism goes beyond your name to other arbitrary features of yourself, such as your birthday. In a university study, students were given an essay to read about the Russian monk Rasputin. For half the students, Rasputin's birthday was mentioned in the essay—and it was gimmicked so that it "happened" to be the same as the reader's own birthday. For the other half of the students, a birthday different from their own was used; otherwise the essays were identical. At the end of the reading, the students were asked to answer several questions covering what they thought of Rasputin as a person. Those who believed they shared a birthday with Rasputin gave him more generous ratings.[12] They simply liked him more, without having any conscious access as to why.

The magnetic power of unconscious self-love goes beyond what and whom you prefer. Incredibly, it can subtly influence where you live and what you do, as well. Psychologist Brett Pelham and his colleagues plumbed public records and found that people with birthdays on February 2 (2/2) are disproportionately likely to move to cities with a reference to the number two in their names, such as Twin Lakes, Wisconsin. People born on 3/3 are statistically overrepresented in places like Three Forks, Montana, as are people born on 6/6 in places like Six Mile, South Carolina, and so on for all the birthdays and cities the authors could find. Consider how amazing that is: associations with the numbers in people's arbitrary birth dates can be influential enough to sway their residential choices, however slightly. Again, it's unconscious.

Implicit egotism can also influence what you chose to do with your life. By analyzing professional membership directories, Pelham and his colleagues found that people named Denise or Dennis are disproportionately likely to become dentists, while people named Laura or Lawrence are more likely to become lawyers, and people with names like George or Georgina to become geologists. They also found that owners of roofing companies are more likely to have a first initial of R instead of H, while hardware store owners are more likely to have names beginning with H instead of R.[13] A different study mined freely available online professional databases to find that physicians have disproportionately more surnames that include *doc*, *dok*, or *med*, while lawyers are more likely to have *law*, *lau*, or *att* in their surnames.[14]

As crazy as it sounds, all these findings passed the statistical thresholds for significance. The effects are not large, but they're verifiable. We are influenced by drives to which we have little access, and which we never would have believed had not the statistics laid them bare.

TICKLING THE BRAIN BELOW THE SURFACE OF AWARENESS

Your brain can be subtly manipulated in ways that change your future behavior. Imagine I ask you to read some pages of text. Later, I ask you to fill in the blanks of some partial terms, such as *chi___ se___*. You're more likely to choose terms that you've recently seen—say, *chicken sexer* rather than *china set*—whether or not you have any explicit memory of having recently seen those words.[15] Similarly, if I ask you to fill in the blanks in some word, such as *s_bl_m_na_*, you are better able to do so if you've previously seen the word on a list, whether or not you remember having seen it.[16] Some part of your brain has been touched and changed by the words on the list. This effect is called priming: your brain has been primed like a pump.[17]

Priming underscores the point that implicit memory systems are fundamentally separate from explicit memory systems: even when the second one has lost the data, the former one has a lock on it. The separability between the systems is again illustrated by patients with anterograde amnesia resulting from brain damage. Severely amnesic patients can be primed to fill in partial words even though they have no conscious recollection of having been presented with any text in the first place.[18]

Beyond a temporary tickling of the brain, the effects of previous exposure can be long lasting. If you have seen a picture of someone's face before, you will judge them to be more attractive upon a later viewing. This is true even when you have no recollection of ever having seen them previously.[19] This is known as the *mere exposure effect*, and it illustrates the worrisome fact that your implicit memory influences your interpretation of the world—which things you like, don't like, and so on. It will come as no surprise to you that the mere exposure effect is part of the magic behind product branding, celebrity building, and political campaigning: with repeated exposure to a product or face, you come to prefer it more. The mere exposure effect is why people in the public spotlight are not always as disturbed as one might expect by negative press. As famous personalities often quip, "The only bad publicity is no publicity," or "I don't care what the newspapers say about me as long as they spell my name right."[20]

Another real-world manifestation of implicit memory is known as the *illusion-of-truth effect*: you are more likely to believe that a statement is true if you have heard it before—whether or not it is actually true. In one study, subjects rated the validity of plausible sentences every two weeks. Without letting on, the experimenters snuck in some repeat sentences (both true and false ones) across the testing sessions. And they found a clear result: if subjects had heard a sentence in previous weeks, they were more likely to now rate it as true, even if they swore they had never heard it before.[21] This is the case even when the experimenter *tells* the subjects that the sentences they are about to hear are false: despite this, mere exposure to an idea is enough to boost its believability upon later contact.[22] The illusion-of-truth effect highlights the potential danger for people who are repeatedly exposed to the same religious edicts or political slogans.

A simple pairing of concepts can be enough to induce an unconscious association and, eventually, the sense that there is something familiar and true about the pairing. This is the basis of every ad we've ever seen that pairs a product with attractive, cheery, and sexually charged people. And it's also the basis of a move made by George W. Bush's advertising team during his 2000 campaign against Al Gore. In Bush's $2.5 million dollar television commercial, a frame with the word RATS flashes on the screen in conjunction with "The Gore prescription plan." In the next moment it becomes clear that the word is actually the end of the word BUREAU-CRATS, but the effect the ad makers were going for was obvious—and, they hoped, memorable.

THE HUNCH

Imagine that you arrange all your fingers over ten buttons, and each button corresponds to a colored light. Your task is simple: each time a light blinks on, you hit the corresponding button as quickly as you can. If the sequence of lights is random, your reaction times will generally not be very fast; however, investigators discovered that if there is a hidden pattern to the lights, your reaction times will eventually speed up, indicating that you have picked up on the sequence and can make some sort of predictions about which light will flash next. If an unexpected light then comes on, your reaction time will be slow again. The surprise is that this speed up works even when you are completely unaware of the sequence; the conscious mind does not need to be involved at all for this type of learning to occur.[23] Your ability to name what is going to occur next is limited or non-existent. And yet you might have a *hunch*.

Sometimes these things can reach conscious awareness, but not always—and when they do, they do so slowly. In 1997, neuro-scientist Antoine Bechara and his colleagues laid out four decks of cards in front of subjects and asked them to choose one card at a time. Each card revealed a gain or loss of money. With time, the subjects began to realize that each deck had a character to it: two of the decks were "good," meaning that the subjects would make money, while the other two were "bad," meaning they would end up with a net loss.

As subjects pondered which deck to draw from, they were stopped at various points by the investigators and asked for their opinion: Which decks were good? Which were bad? In this way, the investigators found that it typically required about twenty-five draws from the decks for subjects to be able to say which ones they thought were good and bad. Not terribly interesting, right? Well, not yet.

The investigators also measured the subject's skin conductance response, which reflects the activity of the autonomic (fight-or-flight) nervous system. And here they noticed something amazing: the autonomic nervous system picked up on the statistics of the deck well before a subject's consciousness did. That is, when subjects reached for the bad decks, there was an anticipatory spike of activity—essentially, a warning sign.[24] This spike was detectable by about the thirteenth card draw. So *some* part of the subjects' brains was picking up on the expected return from the decks well before the subjects' conscious minds could access that information. And the information was being delivered in the form of a "hunch": subjects began to choose the good decks even before they could consciously say why. This means that conscious knowledge of the situation was not required for making advantageous decisions.

Even better, it turned out that people *needed* the gut feeling: without it their decision making would never be very good. Damasio and his colleagues ran the card-choice task using patients with damage to a frontal part of the brain called the ventromedial prefrontal cortex, an area involved in making decisions. The team discovered that these patients were unable to form the anticipatory warning signal of the galvanic skin response. The patients' brains simply weren't picking up on the statistics and giving them an admonition. Amazingly, even after these patients consciously realized which decks were bad, they *still* continued to make the wrong choices. In other words, the gut feeling was essential for advantageous decision making.

This led Damasio to propose that the feelings produced by physical states of the body come to guide behavior and decision making.[25] Body states become linked to outcomes of events in the world. When something bad happens, the brain leverages the entire body (heart rate, contraction of the gut, weakness of the muscles, and so on) to register that feeling, and that feeling becomes associated with the event. When the event is next pondered, the brain essentially runs a simulation, reliving the physical feelings of the event. Those feelings then serve to navigate, or at least bias, subsequent decision making. If the feelings from a given event are bad, they dissuade the action; if they are good, they encourage it.

In this view, physical states of the body provide the hunches that can steer behavior. These hunches turn out to be correct more often than chance would predict, mostly because your unconscious brain is picking up on things first, and your consciousness lags behind.

In fact, conscious systems can break entirely, with no effect on the unconscious systems. People with a condition called prosopagnosia cannot distinguish between familiar and unfamiliar faces. They rely entirely on cues such as hairlines, gait, and voices to recognize people they know. Pondering this condition led researchers Daniel Tranel and Antonio Damasio to try something clever: even though prosopagnosics cannot consciously recognize faces, would they have a measurable skin conductance response to faces that were familiar? Indeed, they did. Even though the prosopagnosic truly insists on being unable to recognize faces, *some* part of his brain can (and does) distinguish familiar faces from unfamiliar ones.

If you cannot always elicit a straight answer from the unconscious brain, how can you access its knowledge? Sometimes the trick is merely to probe what your gut is telling you. So the next time a friend laments that she cannot decide between two options, tell her the easiest way to solve her problem: flip a coin. She should specify which option belongs to heads and which to tails, and then let the coin fly. The important part is to assess her gut feeling after the coin lands. If she feels a subtle sense of relief at being "told" what to do by the coin, that's the right choice for her. If, instead, she concludes that it's ludicrous for her to make a decision based on a coin toss, that will cue her to choose the other option.

*   *   *

So far we've been looking at the vast and sophisticated knowledge that lives under the surface of awareness. We've seen that you don't have access to the details of how your brain does things, from reading letters to changing lanes. So what role does the conscious mind play, if any, in all your know-how? A big one, it turns out—because much of the knowledge stored in the depths of the unconscious brain began life in the form of conscious plans. We turn to this now.

THE ROBOT THAT WON WIMBLEDON

Imagine that you have risen through the ranks to the top tennis tournament in the world and you are now poised on a green court facing the planet's greatest tennis robot. This robot has incredibly miniaturized components and self-repairing parts, and it runs on such optimized energy principles that it can consume three hundred grams of hydrocarbons and then leap all over the court like a mountain goat. Sounds like a formidable opponent, right? Welcome to Wimbledon—you're playing against a human being.

The competitors at Wimbledon are rapid, efficient machines that play tennis shockingly well. They can track a ball traveling ninety miles per hour, move toward it rapidly, and orient a small surface to intersect its trajectory. And these professional tennis players do almost none of this consciously. In exactly the same way that you read letters on a page or change lanes, they rely entirely on their unconscious machinery. They are, for all practical purposes, robots. Indeed, when Ilie Nastase lost the Wimbledon final in 1976, he sullenly said of his winning opponent, Björn Borg, "He's a robot from outer space."

But these robots are *trained by* conscious minds. An aspiring tennis player does not have to know anything about building robotics (that was taken care of by evolution). Rather, the challenge is to *program* the robotics. In this case, the challenge is to program the machinery to devote its flexible computational resources to rapidly and accurately volleying a fuzzy yellow ball over a short net.

And this is where consciousness plays a role. Conscious parts of the brain train other parts of the neural machinery, establishing the goals and allocating the resources. "Grip the racket lower when you swing," the coach says, and the young player mumbles that to herself. She practices her swing over and over, thousands of times, each time setting as her end point the goal of smashing the ball directly into the other quadrant. As she serves again and again, the robotic system makes tiny adjustments across a network of innumerable synaptic connections. Her coach gives feedback which she needs to hear and understand consciously. And she continually incorporates the instructions ("Straighten your wrist. Step into the swing.") into the training of the robot until the movements become so ingrained as to no longer be accessible.

Consciousness is the long-term planner, the CEO of the company, while most of the day-to-day operations are run by all those parts of her brain to which she has no access. Imagine a CEO who has inherited a giant blue-chip company: he has some influence, but he is also coming into a situation that has already been evolving for a long time before he got there. His job is to define a vision and make long-term plans for the company, insofar as the technology of the company is able to support his policies. This is what consciousness does: it sets the goals, and the rest of the system learns how to meet them.

You may not be a professional tennis player, but you've been through this process if you ever learned to ride a bicycle. The first time you got on, you wobbled and crashed and tried desperately to figure it out. Your conscious mind was heavily involved. Eventually, after an adult guided the bicycle along, you became able to ride on your own. After some time, the skill became like a reflex. It became automatized. It became just like reading and speaking your language, or tying your shoes, or recognizing your father's walk. The details became no longer conscious and no longer accessible.

One of the most impressive features of brains—and especially human brains—is the flexibility to learn almost any kind of task that comes its way. Give an apprentice the desire to impress his master in a chicken-sexing task, and his brain devotes its massive resources to distinguishing males from females. Give an unemployed aviation enthusiast a chance to be a national hero, and his brain learns to distinguish enemy aircraft from local flyboys. This flexibility of learning accounts for a large part of what we consider human intelligence. While many animals are properly called intelligent, humans distinguish themselves in that they are so *flexibly* intelligent, fashioning their neural circuits to match the tasks at hand. It is for this reason that we can colonize every region on the planet, learn the local language we're born into, and master skills as diverse as playing the violin, high-jumping and operating space shuttle cockpits.

MANTRA OF THE FAST AND EFFICIENT BRAIN: BURN JOBS INTO THE CIRCUITRY

When the brain finds a task it needs to solve, it rewires its own circuitry until it can accomplish the task with maximum efficiency.[26] The task becomes burned into the machinery. This clever tactic accomplishes two things of chief importance for survival.

The first is *speed*. Automatization permits fast decision making. Only when the slow system of consciousness is pushed to the back of the queue can rapid programs do their work. Should I swing forehand or backhand at the approaching tennis ball? With a ninety-mile-per-hour projectile on its way, one does not want to cognitively slog through the different options. A common misconception is that professional athletes can see the court in "slow motion," as suggested by their rapid and smooth decision making. But automatization simply allows the athletes to anticipate relevant events and proficiently decide what to do. Think about the first time you tried a new sport. More-experienced players defeated you with the most elementary moves because you were struggling with a barrage of new information—legs and arms and jumping bodies. With experience, you learned which twitches and feints were the important ones. With time and automatization, you achieved speed both in deciding and in acting.

The second reason to burn tasks into the circuitry is *energy efficiency*. By optimizing its machinery, the brain minimizes the energy required to solve problems. Because we are mobile creatures that run on batteries, energy saving is of the highest importance.[27] In his book *Your Brain Is (Almost) Perfect*, neuroscientist Read Montague highlights the impressive energy efficiency of the brain, comparing chess champion Garry Kasparov's energy usage of about 20 watts to the consumption of his computerized competitor Deep Blue, in the range of thousands of watts. Montague points out that Kasparov played the game at normal body temperature, while Deep Blue was burning hot to the touch and required a large collection of fans to dissipate the heat. Human brains run with superlative efficiency.

Kasparov's brain is so low-powered because Kasparov has spent a lifetime burning chess strategies into economical rote algorithms. When he started playing chess as a boy, he had to walk himself through cognitive strategies about what to do next—but these were highly inefficient, like the moves of an overthinking, second-guessing tennis player. As Kasparov improved, he no longer had to consciously walk through the unfolding steps of a game: he could perceive the chess board rapidly, efficiently, and with less conscious interference.

In one study on efficiency, researchers used brain imaging while people learned how to play the video game Tetris. The subjects' brains were highly active, burning energy at a massive scale while the neural networks searched for the underlying structures and strategies of the game. By the time the subjects became experts at the game, after a week or so, their brains consumed very little energy while playing. It's not that the player became better despite the brain being quieter; the player became better *because* the brain was quieter. In these players, the skills of Tetris has been burned down into the circuitry of the system, such that there were now specialized and efficient programs to deal with it.

As an analogy, imagine a warring society that suddenly finds itself with no more battles to wage. Its soldiers decide to turn to agriculture. At first they use their battle swords to dig little holes for seeds—a workable but massively inefficient approach. After a time, they beat their swords into plowshares. They optimize their machinery to meet the task demands. Just like the brain, they've modified what they have to address the task at hand.

This trick of burning tasks into the circuitry is fundamental to how brains operate: they change the circuit board of their machinery to mold themselves to their mission. This allows a difficult task that could be accomplished only clumsily to be achieved with rapidity and efficiency. In the logic of the brain, if you don't have the right tool for the job, *create it*.

*   *   *

So far we've learned that consciousness tends to interfere with most tasks (remember the unhappy centipede in the ditch)—but it *can* be helpful when setting goals and training the robot. Evolutionary selection has presumably tuned the exact amount of access the conscious mind has: too little, and the company has no direction; too much, and the system gets bogged down solving problems in a slow, clunky, energy-inefficient manner.

When athletes make mistakes, coaches typically yell, "*Think out there!*" The irony is that a professional athlete's goal is to *not* think. The goal is to invest thousands of hours of training so that in the heat of the battle the right maneuvers will come automatically, with no interference from consciousness. The skills need to be pushed down into the players' circuitry. When athletes "get into the zone," their well-trained unconscious machinery runs the show, rapidly and efficiently. Imagine a basketball player standing at the free-throw line. The crowd yells and stomps to distract him. If he's running on conscious machinery, he's certain to miss. Only by relying on the overtrained, robotic machinery can he hope to drain the ball through the basket.[28]

Now you can leverage the knowledge gained in this chapter to always win at tennis. When you are losing, simply ask your opponent how she serves the ball so well. Once she contemplates the mechanics of her serve and tries to explain it, she's sunk.

We have learned that the more things get automatized, the less conscious access we have. But we're just getting started. In the next chapter we'll see how information can get buried even deeper.

4

The Kinds of Thoughts
That Are Thinkable

"Man is a plant which bears thoughts, just as a rose-tree
bears roses and an apple-tree bears apples."
—Antoine Fabre D'Olivet,
L'Histoire philosophique du genre humain

Spend a moment thinking about the most beautiful person you
know. It would seem impossible for eyes to gaze upon this person
and not be intoxicated with attraction. But everything depends on
the evolutionary program those eyes are connected to. If the eyes
belong to a frog, this person can stand in front of it all day—even
naked—and will attract no attention, perhaps only a bit of suspicion. And the lack of interest is mutual: humans are attracted to
humans, frogs to frogs.

Nothing seems more natural than desire, but the first thing to
notice is that we're wired only for species-appropriate desire. This
underscores a simple but crucial point: the brain's circuits are
designed to generate behavior that is appropriate to our survival.
Apples and eggs and potatoes taste good to us not because the
shapes of their molecules are inherently wonderful, but because
they're perfect little packages of sugars and proteins: energy dollars
you can store in your bank. Because those foods are useful, we
are engineered to find them tasty. Because fecal matter contains
harmful microbes, we have developed a hardwired aversion to
eating it. Note that baby koalas—known as joeys—eat their
mother's fecal matter to obtain the right bacteria for their diges-
tive systems. These bacteria are necessary for the joeys to survive
on otherwise-poisonous eucalyptus leaves. If I had to guess, I'd say

that fecal matter tastes as delicious to the joey as an apple does to you. Nothing is inherently tasty or repulsive—it depends on your needs. Deliciousness is simply an index of usefulness.

Many people are already familiar with these concepts of attraction or tastiness, but it is often difficult to appreciate how deep this evolutionary carving goes. It's not simply that you are attracted to humans over frogs or that you like apples more than fecal matter—these same principles of hardwired thought guidance apply to all of your deeply held beliefs about logic, economics, ethics, emotions, beauty, social interactions, love, and the rest of your vast mental landscape. Our evolutionary goals navigate and structure our thoughts. Chew on that for a moment. It means there are certain kinds of thoughts we *can* think, and whole categories of thoughts we cannot. Let's begin with all the thoughts you didn't even know you were missing.

THE UMWELT: LIFE ON THE THIN SLICE

"Incredible the Lodging
But limited the Guest."
—Emily Dickinson

In 1670, Blaise Pascal noted with awe that "man is equally incapable of seeing the nothingness from which he emerges and the infinity in which he is engulfed."[1] Pascal recognized that we spend our lives on a thin slice between the unimaginably small scales of the atoms that compose us and the infinitely large scales of galaxies.

But Pascal didn't know the half of it. Forget atoms and galaxies— we can't even see most of the action at our *own* spatial scales. Take what we call visible light. We have specialized receptors in the backs of our eyes that are optimized for capturing the electromagnetic radiation that bounces off objects. When these receptors catch some radiation, they launch a salvo of signals into the brain. But we do not perceive the *entire* electromagnetic spectrum, only a part of it. The part of the light spectrum that is visible to us is less than a ten-trillionth of it. The rest of the spectrum—carrying TV shows, radio signals, microwaves, X-rays, gamma rays, cell phone conversations, and so on—flows through us with no awareness on our part.[2] CNN news is passing through your body right now and you are utterly blind to it, because you have no specialized receptors for that part of the spectrum. Honeybees, by contrast, include information carried on ultraviolet wavelengths in their reality, and rattlesnakes include infrared in their view of the world. Machines in the hospital see the X-ray range, and machines in the dashboard of your car see the radio frequency range. But you can't sense any of these. Even though it's the same "stuff"—electromagnetic radiation—you don't come equipped with the proper sensors. No matter how hard you try, you're not going to pick up signals in the rest of the range.

What you are able to experience is completely limited by your biology. This differs from the commonsense view that our eyes, ears, and fingers passively receive an objective physical world outside of ourselves. As science marches forward with machines that can see what we can't, it has become clear that our brains sample just a small bit of the surrounding physical world. In 1909, the Baltic German biologist Jakob von Uexküll began to notice that different animals in the same ecosystem pick up on different signals from their environment.[3] In the blind and deaf world of the tick, the important signals are temperature and the odor of butyric acid. For the black ghost knifefish, it's electrical fields. For the echolocating bat, air-compression waves. So von Uexküll introduced a new concept: the part that you are able to see is known as the *umwelt* (the environment, or surrounding world), and the bigger reality (if there is such a thing) is known as the *umgebung*.

Each organism has its own umwelt, which it presumably assumes to be the entire objective reality "out there." Why would we ever stop to think that there is more beyond what we can sense? In the movie *The Truman Show*, the eponymous Truman lives in a world completely constructed around him (often on the fly) by an intrepid television producer. At one point an interviewer asks the producer, "Why do you think Truman has never come close to discovering the true nature of his world?" The producer replies, "We accept the reality of the world with which we're presented." He hit the nail on the head. We accept the umwelt and stop there.

Ask yourself what it would be like to have been blind from birth. Really think about this for a moment. If your guess is "it would something like blackness" or "something like a dark hole where vision should be," you're wrong. To understand why, imagine you're a scent dog such as a bloodhound. Your long nose houses two hundred million scent receptors. On the outside, your wet nostrils attract and trap scent molecules. The slits at the corners of each nostril flare out to allow more air flow as you sniff. Even your floppy ears drag along the ground and kick up scent molecules. Your world is all about smelling. One afternoon, as you're following your master, you stop in your tracks with a revelation. What is it like to have the pitiful, impoverished nose of a human being? What can humans possibly detect when they take in a feeble little noseful of air? Do they suffer a blackness? A hole of smell where smell is supposed to be?

Because you're a human, you know the answer is no. There is no hole or blackness or missing feeling where the scent is absent. You accept your reality as it's presented to you. Because you don't have the smelling capabilities of a bloodhound, it doesn't even strike you that things could be different. The same goes for people with color blindness: until they learn that others can see hues they cannot, the thought does not even hit their radar screen.

If you are not color-blind, you may well find it difficult to imagine yourself as color-blind. But recall what we learned earlier: that some people see *more* colors than you do. A fraction of women have not just three but four types of color photoreceptors—and as a result they can distinguish colors that the majority of humankind will never differentiate.[4] If you are not a member of that small female population, then you have just discovered something about your own impoverishments that you were unaware of. You may not have thought of yourself as color-blind, but to those ladies supersensitive to hues, you are. In the end, it does not ruin your day; instead, it only makes you wonder how someone else can see the world so strangely.

And so it goes for the congenitally blind. They are not missing anything; they do not see blackness where vision is missing. Vision was never part of their reality in the first place, and they miss it only as much as you miss the extra scents of the bloodhound dog or the extra colors of the tetrachromatic women.

*   *   *

There is a large difference between the umwelts of humans and those of ticks and bloodhounds, but there can even be quite a bit of individual variability between humans. Most people, during some late-night departure from quotidian thinking, ask their friends the following sort of question: How do I know that what I experience as red and what you experience as red is the same thing? This is a good question, because as long as we agree on labeling some feature "red" in the outside world, it doesn't matter if the swatch experienced by you is what I internally perceive as canary yellow. I call it red, you call it red, and we can appropriately transact over a hand of poker.

But the problem actually runs deeper. What I call vision and what you call vision might be different—mine might be upside down compared to yours, and we would never know. And it wouldn't matter, as long as we agree on what to call things and how to point to them and where to navigate in the outside world.

This sort of question used to live in the realm of philosophical speculation, but it has now been promoted to the realm of scientific experiment. After all, across the population there are slight differences in brain function, and sometimes these translate directly into different ways of experiencing the world. And each individual believes his way is *reality*. To get a sense of this, imagine a world of magenta Tuesdays, tastes that have shapes, and wavy green symphonies. One in a hundred otherwise normal people experience the world this way, because of a condition called synesthesia (meaning "joined sensation").[5] In synesthetes, stimulation of a sense triggers an anomalous sensory experience: one may hear colors, taste shapes, or systematically experience other sensory blendings. For example, a voice or music may not only be heard but also seen, tasted, or felt as a touch. Synesthesia is a fusion of different sensory perceptions: the feel of sandpaper might evoke an F-sharp, the taste of chicken might be accompanied by a feeling of pinpoints on the fingertips, or a symphony might be experienced in blues and golds. Synesthetes are so accustomed to the effects that they are surprised to find that others do not share their experiences. These synesthetic experiences are not abnormal in any pathological sense; they are simply unusual in a statistical sense.

Synesthesia comes in many varieties, and having one type gives you a high chance of having a second or third type. Experiencing the days of the week in color is the most common manifestation of synesthesia, followed by colored letters and numbers. Other common varieties include tasted words, colored hearing, number lines perceived as three-dimensional forms, and letters and numerals experienced as having gender and personalities.[6]

Synesthetic perceptions are involuntary, automatic, and consistent over time. The perceptions are typically basic, meaning that what is sensed is something like a simple color, shape, or texture, rather than something pictorial or specific (for example, synesthetes don't say, "This music makes me experience a vase of flowers on a restaurant table").

Why do some people see the world this way? Synesthesia is the result of increased cross talk among sensory areas in the brain. Think of it like neighboring countries with porous borders on the brain's map. And this cross talk results from tiny genetic changes that pass down family lineages. Think about that: microscopic changes in brain wiring can lead to different realities.[7] The mere existence of synesthesia demonstrates that more than one kind of brain—and one kind of mind—is possible.

Let's zoom in on a particular form of synesthesia as an example. For most of us, February and Wednesday do not have any particular place in space. But some synesthetes experience precise locations in relation to their bodies for numbers, time units, and other concepts involving sequence or ordinality. They can point to the spot where the number 32 is, where December floats, or where the year 1966 lies.[8] These objectified three-dimensional sequences are commonly called number forms, although more precisely the phenomenon is called spatial sequence synesthesia.[9] The most common types of spatial sequence synesthesia involve days of the week, months of the year, the counting integers, or years grouped by decade. In addition to these common types, researchers have encountered spatial configurations for shoe and clothing sizes, baseball statistics, historical eras, salaries, TV channels, temperature, and more. Some individuals possess a form for only one sequence; others have forms for more than a dozen. Like all synesthetes, they express amazement that not everyone visualizes sequences the way they do. If you are not synesthetic yourself, the twist is this: it is difficult for synesthetes to understand how people cope *without* a visualization of time. Your reality is as strange to them as theirs is to you. They accept the reality presented to them, as you do yours.[10]

Nonsynesthetes often imagine that sensing extra colors, textures, and spatial configurations would somehow be a perceptual burden: "Doesn't it drive them crazy having to cope with all the extra bits?" some people ask. But the situation is no different from a color-blind person telling a person with normal vision, "You poor thing. Everywhere you look you're always seeing colors. Doesn't it drive you crazy to have to see everything in *colors*?" The answer is that colors do not drive us crazy, because seeing in color is normal to most people and constitutes what we accept as reality. In the same way, synesthetes are not driven crazy by the extra dimensions. They've never known reality to be anything else. Most synesthetes live their entire lives never knowing that others see the world differently than they do.

Synesthesia, in its dozens of varieties, highlights the amazing differences in how individuals subjectively see the world, reminding us that each brain uniquely determines what it perceives, or is capable of perceiving. This fact brings us back to our main point here—namely, that reality is far more subjective than is commonly supposed.[11] Instead of reality being passively recorded by the brain, it is actively constructed by it.

* * *

By analogy to your perception of the world, your mental life is built to range over a certain territory, and it is restricted from the rest. There are thoughts you cannot think. You cannot comprehend the sextillion stars of our universe, nor picture a five-dimensional cube, nor feel attracted to a frog. If these examples seem obvious (*Of course I can't!*), just consider them in analogy to seeing in infrared, or picking up on radio waves, or detecting butyric acid as a tick does. Your "thought umwelt" is a tiny fraction of the "thought umgebung." Let's explore this territory.

The function of this wet computer, the brain, is to generate behavior that is appropriate to the environmental circumstances. Evolution has carefully carved your eyes, internal organs, sexual organs, and so on—and also the character of your thoughts and beliefs. We have not only evolved specialized immune defenses against germs, but we have also developed neural machinery to solve specialized problems that were faced by our hunter-gatherer ancestors over 99 percent of our species' evolutionary history. The field of *evolutionary psychology* explores why we think in some ways and not others. While neuroscientists study the pieces and parts that make up brains, evolutionary psychologists study the software that solves social problems. In this view, the physical structure of the brain embodies a set of programs, and the programs are there because they solved a particular problem in the past. New design features are added to or discarded from the species based on their consequences.

Charles Darwin foretold this discipline in the closing of *The Origin of Species*: "In the distant future I see open fields for far more important researches. Psychology will be based on a new foundation, that of the necessary acquirement of each mental power and capacity by gradation." In other words, our psyches evolve, just like eyes and thumbs and wings.

Consider babies. Babies at birth are not blank slates. Instead, they inherit a great deal of problem-solving equipment and arrive at many problems with solutions already at hand.[12] This idea was first speculated about by Darwin (also in *The Origin of Species*), and later carried forward by William James in *The Principles of Psychology*. The concept was then ignored through most of the twentieth century. But it turned out to be right. Babies, helpless as they are, pop into the world with neural programs specialized for reasoning about objects, physical causality, numbers, the biological world, the beliefs and motivations of other individuals, and social interactions. For example, a newborn's brain *expects* faces: even when they are less than ten minutes old, babies will turn toward face-like patterns, but not to scrambled versions of the same pattern.[13] By two and a half months, an infant will express surprise if a solid object appears to pass through another object, or if an object seems to disappear, as though by magic, from behind a screen. Infants show a difference in the way they treat animate versus inanimate objects, making the assumption that animate toys have internal states (intentions) that they cannot see. They also make assumptions about the intentions of adults. If an adult tries to demonstrate how to do something, a baby will impersonate him. But if the adult appears to mess up the demonstration (perhaps punctuated with a "Whoops!") the infant will not try to impersonate what she saw, but instead what she believes the adult intended.[14] In other words, by the time babies are old enough to be tested, they are already making assumptions about the workings of the world.

So although children learn by imitating what is around them— aping their parents, pets and the TV—they are not blank slates.

Take babbling. Deaf children babble in the same way that hearing children do, and children in different countries sound similar even though they are exposed to radically different languages. So the initial babbling is inherited as a preprogrammed trait in humans.

Another example of preprogramming is the so-called mind-reading system—this is the collection of mechanisms by which we use the direction and movement of other people's eyes to infer what they want, know, and believe. For example, if someone abruptly looks over your left shoulder, you'll immediately suppose there is something interesting going on behind you. Our gaze-reading system is fully in place early in infancy. In conditions like autism this system can be impaired. On the flip side, it can be spared even while other systems are damaged, as in a disorder called Williams syndrome, in which gaze reading is fine but social cognition is broadly deficient in other ways.

Prepackaged software can circumvent the explosion of possibilities that a blank-slate brain would immediately run up against. A system that begins with a blank slate would be unable to learn all the complex rules of the world with only the impoverished input that babies receive.[15] It would have to try everything, and it would fail. We know this, if for no other reason, than from the long history of failure of artificial neural networks that start off knowledge-free and attempt to learn the rules of the world.

Our preprogramming is deeply involved in social exchange—the way humans interact with one another. Social interaction has been critical to our species for millions of years, and as a result the social programs have worked their way deep down into the neural circuitry. As the psychologists Leda Cosmides and John Tooby put it, "The heartbeat is universal because the organ that generates it is everywhere the same. This is a parsimonious explanation for the universality of social exchange as well." In other words, the brain, like the heart, doesn't require a particular culture in order to express social behavior—that program comes pre-bundled with the hardware.

Let's turn to a particular example: your brain has trouble with certain types of calculations that it did not evolve to solve, but has an easy time with calculations that involve social issues. Say I have a set of cards with colors on one side and numbers on the other. I deal the cards below and assert the following claim: cards with *even* numbers will always have *primary* colors on their opposite face. Which two cards do you need to turn over to assess whether I'm telling you the truth?

Don't worry if this problem gives you trouble: it's difficult. The answer is that you need to turn over only the number 8 card and the Purple card. If you had turned over the 5 card and found Red on the other side, that would tell you nothing about the truth of the rule, because I made a statement only about even-numbered cards. Likewise, if you'd turned over the Red card and found an odd number on the other side, it would also have no bearing on the logical rule I gave you, because I never specified what odd numbers may have on their other side.

If your brain were wired up for the rules of conditional logic, you would have no problem with this task. But less than a quarter of people get it right, and that's true even if they've had formal training in logic.[16] The fact that the problem is found to be difficult indicates that our brains aren't wired for general logic problems of this sort. Presumably this is because we have gotten by decently well as a species without needing to nail these sorts of logic puzzles.

But here's the twist to the story. If the exact same logic problem is presented in a way that we are hardwired to understand—that is, cast in the vocabulary of things a social human brain cares about—then it is solved easily.[17] Suppose the new rule is this: If you're under 18, you cannot drink alcohol. Now each card, as shown below, has the age of a person on one side and the drink they're holding on the other.

Which cards do you need to turn over to see if the rule is being broken? Here, most participants get it right (the 16 and Tequila cards). Note that the two puzzles are formally equivalent. So why did you find the first one difficult and the second one easier? Cosmides and Tooby argue that the performance boost in the second case represents a neural specialization. The brain cares about social interaction so much that it has evolved special programs devoted to it: primitive functions to deal with issues of entitlement and obligation. In other words, your psychology has evolved to solve social problems such as detecting cheaters—but not to be smart and logical in general.

MANTRA OF THE EVOLVING BRAIN: BURN REALLY GOOD PROGRAMS ALL THE WAY DOWN TO THE DNA

"In general, we're least aware of what our minds do best."
—Marvin Minsky, *The Society of Mind*

Instincts are complex, inborn behaviors that do not have to be learned. They unpack themselves more or less independently of experience. Consider the birth of a horse: it drops out of the mother's womb, rights itself onto its skinny, uncertain legs, wobbles around for a bit, and finally begins to walk and run, following the rest of the herd in a matter of minutes to hours. The foal is not learning to use its legs from years of trial and error, as a human infant does. Instead, the complex motor action is instinctual.

Because of specialized neural circuits that come as standard equipment with brains, frogs are mad with desire for other frogs and cannot imagine what it would mean for a human to command sex appeal—and vice versa. The programs of instinct, carved by the pressures of evolution, keep our behaviors running smoothly and steer our cognition with a firm hand.

Instincts are traditionally thought to be the opposite of reasoning and learning. If you're like most people, you'll consider your dog to operate largely on instincts, while humans appear to run on something other than instincts—something more like *reason*. The great nineteenth-century psychologist William James was the first to get suspicious of this story. And not just suspicious: he thought it was dead wrong. He suggested instead that human behavior may be more flexibly intelligent than that of other animals because we possess *more* instincts than they do, not fewer. These instincts are tools in the toolbox, and the more you have, the more adaptable you can be.

We tend to be blind to the existence of these instincts precisely because they work so well, processing information effortlessly and automatically. Just like the unconscious software of the chicken sexers or plane spotters or tennis players, the programs are burned down so deeply into the circuitry that we can no longer access them. Collectively, these instincts form what we think of as human nature.[18]

Instincts differ from our automatized behaviors (typing, bicycle riding, serving a tennis ball) in that we didn't have to learn them in our lifetime. We inherited them. Our innate behaviors represent ideas so useful that they became encoded into the tiny, cryptic language of DNA. This was accomplished by natural selection over millions of years: those who possessed instincts that favored survival and reproduction tended to multiply.

The key point here is that the specialized, optimized circuitry of instinct confers all the benefits of speed and energy efficiency, but at the cost of being further away from the reach of conscious access. As a result, we have as little access to our hardwired cognitive programs as we do to our tennis serve. This situation leads to what Cosmides and Tooby call "instinct blindness": we are not able to see the instincts that are the very engines of our behavior.[19] These programs are inaccessible to us *not* because they are unimportant, but because they're *critical*. Conscious meddling would do nothing to improve them.

William James realized the hidden nature of instincts and suggested that we coax instincts into the light by a simple mental exercise: try to make the "natural seem strange" by asking "the why of any instinctive human act":

> Why do we smile, when pleased, and not scowl? Why are we unable to talk to a crowd as we talk to a single friend? Why does a particular maiden turn our wits so upside-down? The common man can only say, *Of course* we smile, *of course* our heart palpitates at the sight of the crowd, *of course* we love the maiden, that beautiful soul clad in that perfect form, so palpably and flagrantly made for all eternity to be loved!
>
> And so, probably, does each animal feel about the particular things it tends to do in the presence of particular objects. . . . To the lion it is the lioness which is made to be loved; to the bear, the she-bear. To the broody hen the notion would probably seem monstrous that there should be a creature in the world to whom a nestful of eggs was not the utterly fascinating and precious and never-to-be-too-much-sat-upon object which it is to her.
>
> Thus we may be sure that, however mysterious some animals' instincts may appear to us, our instincts will appear no less mysterious to them.[20]

Our most hardwired instincts have usually been left out of the spotlight of inquiry as psychologists have instead sought to understand uniquely human acts (such as higher cognition) or how things go wrong (such as mental disorders). But the most automatic, effortless acts—those that require the most specialized and complex neural circuitry—have been in front of us all along: sexual attraction, fearing the dark, empathizing, arguing, becoming jealous, seeking fairness, finding solutions, avoiding incest, recognizing facial expressions. The vast networks of neurons underpinning these acts are so well tuned that we fail to be aware of their normal operation. And just as it was for the chicken sexers, introspection is useless for accessing programs burned into the circuitry. Our conscious assessment of an activity as easy or natural can lead us to grossly underestimate the complexity of the circuits that make it possible. Easy things are hard: most of what we take for granted is neurally complex.

As one illustration of this, consider what has happened in the field of artificial intelligence. In the 1960s it made rapid progress in programs that could deal with fact-driven knowledge, such as "a horse is a type of mammal." But then the field slowed almost to a halt. It turned out to be much more difficult to crack "simple" problems, such as walking along a sidewalk without falling off the curb, remembering where the cafeteria is, balancing a tall body on two tiny feet, recognizing a friend, or understanding a joke. The things we do rapidly, efficiently, and unconsciously are so difficult to model that they remain unsolved problems.

The more obvious and effortless something seems, the more we need to suspect that it seems that way only because of the massive circuitry living behind it. As we saw in Chapter 2, the act of seeing is so easy and rapid precisely because it is underpinned by complicated, dedicated machinery. The more natural and effortless something *seems*, the less so it *is*.[21] Our lust circuits are not driven by the naked frog because we cannot mate with frogs and they have little to do with our genetic future. On the other hand, as we saw in the first chapter, we *do* care quite a bit about the dilation of a woman's eyes, because this broadcasts important information about sexual interest. We live inside the umwelt of our instincts, and we typically have as little perception of them as the fish does of its water.

BEAUTY: SO PALPABLY AND FLAGRANTLY MADE FOR ALL ETERNITY TO BE LOVED

Why are people attracted to young mates and not to the elderly? Do blondes really have more fun? Why does a briefly glimpsed person appear more attractive than a person at whom we've taken a good look? At this point, you won't be surprised to find that our sense of beauty is burned deeply (and inaccessibly) into the brain—all with the purpose of accomplishing something biologically useful.

Let's return to thinking about the most beautiful person you know. Well-proportioned, effortlessly well liked, magnetic. Our brains are exquisitely honed to pick up on those looks. Simply because of small details of symmetry and structure, that person enjoys a destiny of greater popularity, faster promotions, and a more successful career.

At this point it will not surprise you to discover that our sense of attraction is not something ethereal—properly studied only by the pens of poets—but instead results from specific signals that plug, like a key into a lock, into dedicated neural software.

What people select as beautiful qualities primarily reflect signs of fertility brought on by hormonal changes. Until puberty the faces and body shapes of boys and girls are similar. The rise in estrogen in pubescent girls gives them fuller lips, while testosterone in boys produces a more prominent chin, a larger nose, and a fuller jaw. Estrogen causes the growth of the breasts and buttocks, while testosterone encourages the growth of muscles and broad shoulders. So for a female, full lips, full buttocks, and a narrow waist broadcast a clear message: *I'm full of estrogen and fertile.* For a male, it's the full jaw, stubble, and broad chest. This is what we are programmed to find beautiful. Form reflects function.

Our programs are so ingrained that there is little variation across the population. Researchers (as well as purveyors of pornography) have been able to discern a surprisingly narrow range for the female proportions that males find most attractive: the perfect ratio between the waist and hips usually resides between 0.67 and 0.8. The waist-to-hip ratios of *Playboy* centerfolds has remained at about 0.7 over time, even as their average weight has decreased.[22] Women with a ratio in this range are not only judged by males as more attractive, but are also presumed to be more healthy, humorous, and intelligent.[23] As women become older, their features change in ways that depart from these proportions. Middles thicken, lips thin, breasts sag, and so on, all of which broadcast the signal that they are past peak fertility. Even a male teenager with no biology education will be less attracted to an elderly woman than to a young woman. His circuits have a clear mission (reproduction); his conscious mind receives only the need-to-know headline ("She is attractive, pursue her!") and nothing more.

And the hidden neural programs detect more than fertility. Not all fertile women are equally healthy, and therefore they do not all appear equally attractive. The neuroscientist Vilayanur Ramachandran joked that the quip about men preferring blondes may have a biological seed of truth to it: paler women more easily show signs of disease, while the darker complexions of swarthier women can better disguise their imperfections. More health information allows a better choice, and thus is preferable.[24]

Males are often more visually driven than females, but women are nonetheless subject to the same internal forces; they are drawn by the attractive features that flag the maturity of manhood. An interesting twist is that a woman's preferences can change depending on the time of month: women prefer masculine-looking men when they are ovulating, but when not ovulating they prefer softer features—which presumably flag more social and caring behavior.[25]

Although the programs of seduction and pursuit run largely under the machinery of consciousness, the endgame becomes obvious to everyone. This is why thousands of citizens of rich countries shell out for face-lifts, tummy tucks, implants, liposuction, and Botox. They are working to maintain the keys that unlock the programs in other people's brains.

Not surprisingly, we have almost no direct access to the mechanics of our attractions. Instead, visual information plugs into ancient neural modules that drive our behavior. Recall the experiment in the first chapter: when men ranked the beauty of women's faces, they found the women with dilated eyes more attractive, because dilated eyes signal sexual interest. But the men had no conscious access to their decision-making processes.

In a study in my laboratory, participants viewed brief flashes of photographs of men and women and rated their attractiveness.[26] In a later round they were asked to rate the same photos they had seen before, but this time with as much time as they wanted to examine the photos. The result? Briefly glimpsed people are more beautiful. In other words, if you catch a glimpse of someone rounding a corner or driving past quickly, your perceptual system will tell you they are more beautiful than you would otherwise judge them to be. Men show this misjudgment effect more strongly than women, presumably because men are more visual in assessing attraction. This "glimpse effect" accords with everyday experience, in which a man catches a brief glimpse of a woman and believes he has just missed a rare beauty; then, when he rushes around the corner, he discovers that he was mistaken. The effect is clear, but the reason behind it is not. Why should the visual system, given just a bit of fleeting information, always err on the side of believing that the woman is more beautiful? In the absence of clear data, why wouldn't your perceptual system simply strike for the middle and judge the woman to be average, or even below average?

The answer pivots on the demands of reproduction. If you believe a briefly glimpsed unattractive person is beautiful, it requires only a double take to correct the mistake—not much of a cost. On the other hand, if you mistake an attractive mate for an unattractive one, you can say *sayonara* to a potentially rosy genetic future. So it behooves a perceptual system to serve up the fish tale that a briefly glimpsed person is attractive. As with the other examples, all your conscious brain knows is that you just passed an incredible beauty driving the other way in traffic; you have no access to the neural machinery nor to the evolutionary pressures that manufactured the belief for you.

Concepts learned from experience can also take advantage of these hardwired mechanisms of attraction. In a recent study, researchers tested whether being unconsciously primed for the concept of alcohol would (also unconsciously) tickle the concepts associated with alcohol, such as sex and sexual desire.[27] Men were shown words like *beer* or *bean*—but the words were flashed too rapidly to be consciously perceived. The men then rated the attractiveness of photographs of women. After being unconsciously primed with the alcohol-related words (like *beer*), the subjects rated the photographs as more attractive. And the males who more strongly believed that alcohol increases sexual desire showed the strongest effect.

Attraction is not a fixed concept, but instead adjusts according to the demands of the situation—take, for example, the concept of being in heat. Almost all female mammals give off clear signals when they are in heat. The rear end of female baboons turns bright pink, an unmistakable and irresistible invitation for a lucky male baboon. Human females, on the other hand, are unique in that they participate in mating year-round. They do not broadcast any special signal to publicize when they are fertile.[28]

Or don't they? It turns out that a woman is considered to be most beautiful just at the peak of fertility in her menstrual cycle—about ten days before menses.[29] This is true whether she's judged by men or by women, and it's not a matter of the way she acts: it is perceived even by those looking at her photographs. So her good looks broadcast her level of fertility. Her signals are subtler than the baboon's rear end, but they only need to be clear enough to tickle the dedicated, unconscious machinery of the males in the room. If they can reach those circuits, the mission is accomplished. They also reach the circuitry of other females: women are quite sensitive to the effect of other women's cycles, perhaps because this lets them assess their competitors when fighting for mates. It is not yet clear what the tip-offs for fertility are—they may include some quality of the skin (as tone becomes lighter during ovulation) or the fact that a woman's ears and breasts become more symmetrical in the days leading up to ovulation.[30] Whatever the constellation of clues, our brains are engineered to latch on, even while the conscious mind has no access. The mind merely senses the almighty and inexplicable tug of desire.

The effects of ovulation and beauty are not just assessed in the laboratory—they are measurable in real-life situations. A recent study by scientists in New Mexico counted up the tips made by lap dancers at local strip clubs and correlated this with the menstrual cycles of the dancers.[31] During peak fertility, dancers raked in an average of $68 an hour. When they were menstruating, they earned only about $35. In between, they averaged $52. Although these women were presumably acting in a high capacity of flirtation throughout the month, their change in fertility was broadcast to hopeful customers by changes in body odor, skin, waist-to-hip ratio, and likely their own confidence as well. Interestingly, strippers on birth control did not show any clear peak in performance, and earned only a monthly average of $37 per hour (versus an average of $53 per hour for strippers not on birth control). Presumably they earned less because the pill leads to hormonal changes (and cues) indicative of early pregnancy, and the dancers were thus less interesting to Casanovas in the gentlemen's clubs.

What does this research tell us? It tells us that fiscally concerned strippers should eschew contraception and double up their shifts just before ovulation. More importantly, it drives home the point that the beauty of the maiden (or man) is neurally preordained. We have no conscious access to the programs, and can tease them out only with careful studies. Note that brains are quite good at detecting the subtle cues involved. Returning to the most beautiful person you know, imagine that you measured the distance between his or her eyes, as well as nose length, lip thickness, chin shape, and so on. If you compared these measurements to those of a not-so-attractive person, you would find that the differences are subtle.

To a space alien or a German Shepherd dog, the two humans would be indistinguishable, just as attractive and unattractive space aliens and German Shepherd dogs are difficult for you to tell apart. But the small differences within your own species have a great deal of effect in your brain. As an example, some people find the sight of a woman in short shorts intoxicating and a male in short shorts repulsive, even though the two scenes are hardly different from a geometrical perspective. Our ability to make subtle distinctions is exquisitely fine-grained; our brains are engineered to accomplish the clear-cut tasks of mate selection and pursuit. All of it rides under the surface of conscious awareness—we get to simply enjoy the lovely feelings that bubble up.

*   *   *

Beauty judgments are not only constructed by your visual system but are influenced by smell as well. Odor carries a great deal of information, including information about a potential mate's age, sex, fertility, identity, emotions, and health. The information is carried by a flotilla of drifting molecules. In many animal species, these compounds drive behavior almost entirely; in humans, the information often flies beneath the radar of conscious perception, but nonetheless influences our behavior.

Imagine we give a female mouse a selection of males to mate with. Her choice, far from being random, will be based on the interplay between her genetics and the genetics of her suitors. But how does she have access to that kind of hidden information? All mammals have a set of genes known as the major histocompatibility complex, or MHC; these genes are key players in our immune systems. Given a choice, the mouse will choose a mate with *dissimilar* MHC genes. Mixing up the gene pool is almost always a good idea in biology: it keeps genetic defects to a minimum and leads to a healthy interplay of genes known as hybrid vigor. So finding genetically distant partners is useful. But how do mice, who are largely blind, pull this off? With their noses. An organ inside their noses picks up pheromones, floating chemicals that carry signals through the air—signals about things such as alarm, food trails, sexual readiness, and, in this case, genetic similarity or difference.

Do humans sense and respond to pheromones the way mice do? No one knows for sure, but recent work has found receptors in the lining of the human nose just like those used in pheromonal signaling in mice.[32] It's not clear if our receptors are functional, but the behavioral research is suggestive.[33] In a study at the University of Bern, researchers measured and quantified the MHCs of a group of male and female students.[34] The males were then given cotton T-shirts to wear, so that their daily sweat soaked into the fabric. Later, back in the laboratory, females plunged their noses into the armpits of these T-shirts and picked which body odor they preferred. The result? Exactly like the mice, they preferred the males with more dissimilar MHCs. Apparently our noses are also influencing our choices, again flying the reproduction mission under the radar of consciousness.

Beyond reproduction, human pheromones may also carry invisible signals in other situations. For example, newborns preferentially move toward pads that have been rubbed on their mother's breast rather than clean pads, presumably based on pheromonal cues.[35] And the length of women's menstrual cycles may change after they sniff the armpit sweat of another woman.[36]

Although pheromones clearly carry signals, the degree to which they influence human behavior is unknown. Our cognition is so multilayered that these cues have been reduced to bit players. Whatever other role they have, pheromones serve to remind us that the brain continuously evolves: these molecules unmask the presence of outdated legacy software.

INFIDELITY IN THE GENES?

Consider your attachment to your mother, and the good fortune of her attachment back to you—especially when you needed her as a helpless infant. That sort of bonding is easy enough to imagine as a natural occurrence. But we need merely to scratch the surface to find that social attachment relies on a sophisticated system of chemical signaling. It doesn't happen by default; it happens on purpose. When mice pups are genetically engineered to lack a particular type of receptor in the opioid system (which is involved in pain suppression and reward), they stop caring about separation from their mothers.[37] They let out fewer cries. This is not to say that they are unable to care about things in general—in fact, they are more reactive than normal mice to a threatening male mouse or to cold temperatures. It's simply that they don't seem to bond to their mothers. When they are given a choice between smells from their mother and smells from an unknown mouse, they are just as likely to choose either one. The same thing happens when they are presented with their mother's nest versus a stranger's nest. In other words, pups must be running the proper genetic programs to correctly care about their mothers. This sort of problem may underlie disorders that involve difficulties with attachment, such as autism.

Related to the issue of parental bonding is that of staying faithful to one's partner. Common sense would tell us that monogamy is a decision based on moral character, right? But this leads to the question of what constitutes "character" in the first place. Could this, too, be guided by mechanisms below the radar of consciousness?

Consider the prairie vole. These little creatures dig through shallow underground runways and stay active all year. But unlike other voles and other mammals more generally, prairie voles remain monogamous. They form life-long pair bonds in which they nest together, huddle up, groom, and raise the pups as a team. Why do they show this behavior of committed affiliation while their close cousins are more wanton? The answer pivots on hormones.

When a male vole repeatedly mates with a female, a hormone called vasopressin is released in his brain. The vasopressin binds to receptors in a part of the brain called the nucleus accumbens, and the binding mediates a pleasurable feeling that becomes associated with that female. This locks in the monogamy, which is known as pair-bonding. If you block this hormone, the pair-bonding goes away. Amazingly, when researchers crank up the levels of vasopressin with genetic techniques, they can shift polygamous species to monogamous behavior.[38]

Does vasopressin matter for human relationships? In 2008, a research team at the Karolinska Institute in Sweden examined the gene for the vasopressin receptor in 552 men in long-term heterosexual relationships.[39] The researchers found that a section of the gene called RS3 334 can come in variable numbers: a man might carry no copies of this section, one copy, or two copies. The more copies, the weaker the effect that vasopressin in the bloodstream would have in the brain. The results were surprising in their simplicity. The number of copies correlated with the men's pair-bonding behavior. Men with more copies of RS3 334 scored worse on measures of pair-bonding—including measures of the strength of their relationships, perceived marital problems, and marital quality as perceived by their spouses. Those with two copies were more likely to be unmarried, and if they were married, they were more likely to have marital troubles.

This is not to say that choices and environment don't matter— they do. But it *is* to say that we come into the world with different dispositions. Some men may be genetically inclined to have and hold a single partner, while some may not. In the near future, young women who stay current with the scientific literature may demand genetic tests of their boyfriends to assess how likely they are to make faithful husbands.

Recently, evolutionary psychologists have turned their sights on love and divorce. It didn't take them long to notice that when people fall in love, there's a period of up to three years during which the zeal and infatuation ride at a peak. The internal signals in the body and brain are literally a love drug. And then it begins to decline. From this perspective, we are preprogrammed to lose interest in a sexual partner after the time required to raise a child has passed—which is, on average, about four years.[40] The psychologist Helen Fisher suggests that we are programmed the same way as foxes, who pair-bond for a breeding season, stick around just long enough to raise the offspring, and then split. By researching divorce in nearly sixty countries, Fisher has found that divorce peaks at about four years into a marriage, consistent with her hypothesis.[41] In her view, the internally generated love drug is simply an efficient mechanism to get men and women to stick together long enough to increase the survival likelihood of their young. Two parents are better than one for survival purposes, and the way to provide that safety is to coax them into staying together.

In the same vein, the large eyes and round faces of babies look cute to us not because they possess a natural "cuteness" but because of the evolutionary importance of adults taking care of babies. Those genetic lines that did not find their infants cute no longer exist, because their young were not properly cared for. But survivors like us, whose mental umwelt cannot let us *not* find babies cute, successfully raise babies to compose the next generation.

* * *

We've seen in this chapter that our deepest instincts, as well as the kinds of thoughts we have and even *can* have, are burned into the machinery at a very low level. "This is great news," you might think. "My brain is doing all the right things to survive, and I don't even have to think about them!" True, that is great news. The unexpected part of the news is that the conscious *you* is the smallest bit-player in the brain. It is something like a young monarch who inherits the throne and takes credit for the glory of the country—without ever being aware of the millions of workers who keep the place running.

We'll need some bravery to start considering the limitations of our mental landscape. Returning to the movie *The Truman Show*, at one point an anonymous woman on the telephone suggests to the producer that poor Truman, unwittingly on TV in front of an audience of millions, is less a performer than a prisoner. The producer calmly replies:

> And can you tell me, caller, that you're not a player on the stage of life—playing out your allotted role? He can leave at any time. If his was more than just a vague ambition, if he were absolutely determined to discover the truth, there's no way we could prevent him. I think what really distresses you, caller, is that ultimately Truman prefers the comfort of his "cell," as you call it.

As we begin to explore the stage we're on, we find that there is quite a bit beyond our umwelt. The search is a slow, gradual one, but it engenders a deep sense of awe at the size of the wider production studio.

We're now ready to move one level deeper into the brain, uncovering another layer of secrets about what we've been blithely referring to as *you*, as though you were a single entity.

5

The Brain Is a Team of Rivals

"Do I contradict myself?
Very well then I contradict myself,
(I am large, I contain multitudes.)"
—Walt Whitman, *Song of Myself*

WILL THE TRUE MEL GIBSON PLEASE STAND UP?

On July 28, 2006, the actor Mel Gibson was pulled over for speeding at nearly twice the posted speed limit on the Pacific Coast Highway in Malibu, California. The police officer, James Mee, administered a breathalyzer test, which revealed Gibson's blood alcohol level to be 0.12 percent, well over the legal limit. On the seat next to Gibson sat an open bottle of tequila. The officer announced to Gibson that he was under arrest and asked him to get into the squad car. What distinguished this arrest from other Hollywood inebriations was Gibson's surprising and out-of-place inflammatory remarks. Gibson growled, "Fucking Jews. . . . Jews are responsible for all the wars in the world." He then asked the officer, "Are you a Jew?" Mee was indeed Jewish. Gibson refused to get into the squad car and had to be handcuffed.

Less than nineteen hours later, the celebrity website TMZ.com obtained a leak of the handwritten arrest report and posted it immediately. On July 29, after a vigorous response from the media, Gibson offered a note of apology:

> After drinking alcohol on Thursday night, I did a number of
> things that were very wrong and for which I am ashamed. . . .
> I acted like a person completely out of control when I was

arrested, and said things that I do not believe to be true and which are despicable. I am deeply ashamed of everything I said and I apologize to anyone who I have offended. . . . I disgraced myself and my family with my behavior and for that I am truly sorry. I have battled the disease of alcoholism for all of my adult life and profoundly regret my horrific relapse. I apologize for any behavior unbecoming of me in my inebriated state and have already taken necessary steps to ensure my return to health.

Abraham Foxman, head of the Anti-Defamation League, expressed outrage that there was no reference in the apology to the anti-Semitic slurs. In response, Gibson extended a longer note of contrition specifically toward the Jewish community:

There is no excuse, nor should there be any tolerance, for anyone who thinks or expresses any kind of anti-Semitic remark. I want to apologize specifically to everyone in the Jewish community for the vitriolic and harmful words that I said to a law enforcement officer the night I was arrested on a DUI charge. . . . The tenets of what I profess to believe necessitate that I exercise charity and tolerance as a way of life. Every human being is God's child, and if I wish to honor my God I have to honor his children. But please know from my heart that I am not an anti-Semite. I am not a bigot. Hatred of any kind goes against my faith.

Gibson offered to meet one-on-one with leaders of the Jewish community to "discern the appropriate path for healing." He seemed genuinely contrite, and Abraham Foxman accepted his apology on behalf of the Anti-Defamation League.

Are Gibson's true colors that of an anti-Semite? Or are his true colors those he showed afterward, in his eloquent and apparently heartfelt apologies?

In a *Washington Post* article entitled "Mel Gibson: It Wasn't Just the Tequila Talking," Eugene Robinson wrote, "Well, I'm sorry about his relapse, but I just don't buy the idea that a little tequila, or even a lot of tequila, can somehow turn an unbiased person into a raging anti-Semite—or a racist, or a homophobe, or a bigot of any kind, for that matter. Alcohol removes inhibitions, allowing all kinds of opinions to escape uncensored. But you can't blame alcohol for forming and nurturing those opinions in the first place."

Lending support to that outlook, Mike Yarvitz, the television producer of *Scarborough Country*, drank alcohol on the show until he raised his blood alcohol level to 0.12 percent, Gibson's level that night. Yarvitz reported "not feeling anti-Semitic" after drinking.

Robinson and Yarvitz, like many others, suspected that the alcohol had loosened Gibson's inhibitions and revealed his true self. And the nature of their suspicion has a long history: the Greek poet Alcaeus of Mytilene coined a popular phrase *En oino alétheia* (In wine there is the truth), which was repeated by the Roman Pliny the Elder as *In vino veritas*. The Babylonian Talmud contains a passage in the same spirit: "In came wine, out went a secret." It later advises, "In three things is a man revealed: in his wine goblet, in his purse, and in his wrath." The Roman historian Tacitus claimed that the Germanic peoples always drank alcohol while holding councils to prevent anyone from lying.

But not everyone agreed with the hypothesis that alcohol revealed the true Mel Gibson. The *National Review* writer John Derbyshire argued, "The guy was drunk, for heaven's sake. We all say and do dumb things when we are drunk. If I were to be judged on my drunken escapades and follies, I should be utterly excluded from polite society, and so would you, unless you are some kind of saint." The Jewish conservative activist David Horowitz commented on Fox News, "People deserve compassion when they're in this kind of trouble. I think it would be very ungracious for people to deny it to him." Addiction psychologist G. Alan Marlatt wrote in *USA Today*, "Alcohol is not a truth serum. . . . It may or may not indicate his true feelings."

In fact, Gibson had spent the afternoon before the arrest at the house of a friend, Jewish film producer Dean Devlin. Devlin stated,

"I have been with Mel when he has fallen off, and he becomes a completely different person. It is pretty horrifying." He also stated, "If Mel is an anti-Semite, then he spends a lot of time with us [Devlin and his wife, who is also Jewish], which makes no sense."

So which are Gibson's "true" colors? Those in which he snarls anti-Semitic comments? Or those in which he feels remorse and shame and publicly says, "I am reaching out to the Jewish community for its help"?

Many people prefer a view of human nature that includes a true side and a false side—in other words, humans have a single genuine aim and the rest is decoration, evasion, or cover-up. That's intuitive, but it's incomplete. A study of the brain necessitates a more nuanced view of human nature. As we will see in this chapter, we are made of many neural subpopulations; as Whitman put it, we "contain multitudes." Even though Gibson's detractors will continue to insist that he is truly an anti-Semite, and his defenders will insist that he is not, both may be defending an incomplete story to support their own biases. Is there any reason to believe that it's not possible to have both racist and nonracist parts of the brain?

I AM LARGE, I CONTAIN MULTITUDES

Throughout the 1960s, artificial intelligence pioneers worked late nights to try to build simple robotic programs that could manipulate small blocks of wood: find them, fetch them, stack them in patterns. This was one of those apparently simple problems that turn out to be exceptionally difficult. After all, finding a block of wood requires figuring out which camera pixels correspond to the block and which do not. Recognition of the block shape must be accomplished regardless of the angle and distance of the block. Grabbing it requires visual guidance of graspers that must clench at the correct time, from the correct direction, and with the correct force. Stacking requires an analysis of the rest of the blocks and adjustment to those details. And all these programs need to be coordinated so that they happen at the correct times in the correct sequence. As we have seen in the previous chapters, tasks that appear simple can require great computational complexity.

Confronting this difficult robotics problem a few decades ago, the computer scientist Marvin Minsky and his colleagues introduced a progressive idea: perhaps the robot could solve the problem by distributing the labor among specialized subagents—small computer programs that each bite off a small piece of the problem. One computer program could be in charge of the job *find*. Another could solve the *fetch* problem, and yet another program could take care of *stack block*. These mindless subagents could be connected in a hierarchy, just like a company, and they could report to one another and to their bosses. Because of the hierarchy, *stack block* would not try to start its job until *find* and *fetch* had finished theirs.

This idea of subagents did not solve the problem entirely—but it helped quite a bit. More importantly, it brought into focus a new idea about the working of biological brains. Minsky suggested that human minds may be collections of enormous numbers of machine-like, connected subagents that are themselves mindless.[1] The key idea is that a great number of small, specialized workers can give rise to something like a society, with all its rich properties that no single subagent, alone, possesses. Minsky wrote, "Each mental agent by itself can only do some simple thing that needs no mind or thought at all. Yet when we join these agents in societies—in certain very special ways—this leads to intelligence." In this framework, thousands of little minds are better than one large one.

To appreciate this approach, just consider how factories work: each person on the assembly line is specialized in a single aspect of production. No one knows how to do everything; nor would that equate to efficient production if they did. This is also how government ministries operate: each bureaucrat has one task or a few very specific tasks, and the government succeeds on its ability to distribute the work appropriately. On larger scales, civilizations operate in the same manner: they reach the next level of sophistication when they learn to divide labor, committing some experts to agriculture, some to art, some to warfare, and so on.[2] The division of labor allows specialization and a deeper level of expertise.

The idea of dividing up problems into subroutines ignited the young field of artificial intelligence. Instead of trying to develop a single, all-purpose computer program or robot, computer scientists shifted their goal to equipping the system with smaller "local expert" networks that know how to do one thing, and how to do it well.[3] In such a framework, the larger system needs only to switch which of the experts has control at any given time. The learning challenge now involves not so much how to do each little task but, instead, how to distribute who's doing what when.[4]

As Minsky suggests in his book *The Society of Mind*, perhaps that's all the human brain has to do as well. Echoing William James' concept of instincts, Minsky notes that if brains indeed work this way—as collections of subagents—we would not have any reason to be aware of the specialized processes:

> Thousands and, perhaps, millions of little processes must be involved in how we anticipate, imagine, plan, predict, and prevent—and yet all this proceeds so automatically that we regard it as "ordinary common sense." . . . At first it may seem incredible that our minds could use such intricate machinery and yet be unaware of it.[5]

When scientists began to look into the brains of animals, this society-of-mind idea opened up new ways of looking at things. In the early 1970s, researchers realized that the frog, for example, has at least two separate mechanisms for detecting motion: one system directs the snapping of the frog's tongue to small, darting objects, such as flies, while a second system commands the legs to jump in response to large, looming objects.[6] Presumably, neither of these systems is conscious—instead, they are simple, automated programs burned down into the circuitry.

The society-of-mind framework was an important step forward.

But despite the initial excitement about it, a collection of experts with divided labor has never proven sufficient to yield the properties of the human brain. It is still the case that our smartest robots are less intelligent than a three-year-old child.

So what went wrong? I suggest that a critical factor has been missing from the division-of-labor models, and we turn to that now.

THE DEMOCRACY OF MIND

The missing factor in Minsky's theory was *competition* among experts who all believe they know the right way to solve the problem. Just like a good drama, the human brain runs on conflict.

In an assembly line or government ministry, each worker is an expert in a small task. In contrast, parties in a democracy hold differing opinions *about the same issues*—and the important part of the process is the battle for steering the ship of state. Brains are like representative democracies.[7] They are built of multiple, overlapping experts who weigh in and compete over different choices. As Walt Whitman correctly surmised, we are large and we harbor multitudes within us. And those multitudes are locked in chronic battle.

There is an ongoing conversation among the different factions in your brain, each competing to control the single output channel of your behavior. As a result, you can accomplish the strange feats of arguing with yourself, cursing at yourself, and cajoling yourself to do something—feats that modern computers simply do not do. When the hostess at a party offers chocolate cake, you find yourself on the horns of a dilemma: some parts of your brain have evolved to crave the rich energy source of sugar, and other parts care about the negative consequences, such as the health of your heart or the bulge of your love handles. Part of you wants the cake and part of you tries to muster the fortitude to forgo it. The final vote of the parliament determines which party controls your action—that is, whether you put your hand out or up. In the end, you either eat the chocolate cake or you do not, but you cannot do both.

Because of these internal multitudes, biological creatures can be conflicted. The term *conflicted* could not be sensibly applied to an entity that has a single program. Your car cannot be conflicted about which way to turn: it has one steering wheel commanded by only one driver, and it follows directions without complaint. Brains, on the other hand, can be of two minds, and often many more. We don't know whether to turn toward the cake or away from it, because there are several little sets of hands on the steering wheel of our behavior.

Consider this simple experiment with a laboratory rat: if you put both food *and* an electrical shock at the end of an alley, the rat finds himself stuck at a certain distance from the end. He begins to approach but withdraws; he begins to withdraw but finds the courage to approach again. He oscillates, conflicted.[8] If you outfit the rat with a little harness to measure the force with which he pulls toward food alone and, separately, you measure the force with which he pulls away from an electric shock alone, you find that the rat gets stuck at the point where the two forces are equal and cancel out. The pull matches the push. The perplexed rat has two pair of paws on his steering wheel, each pulling in opposite directions—and as a result he cannot get anywhere.

Brains—whether rat or human—are machines made of conflicting parts. If building a contraption with internal division seems strange, just consider that we already build social machines of this type: think of a jury of peers in a courtroom trial. Twelve strangers with differing opinions are tasked with the single mission of coming to a consensus. The jurors debate, coax, influence, relent—and eventually the jury coheres to reach a single decision. Having differing opinions is not a drawback to the jury system, it is a central feature.

Inspired by this art of consensus building, Abraham Lincoln chose to place adversaries William Seward and Salmon Chase in his presidential cabinet. He was choosing, in the memorable phrase of historian Doris Kearns Goodwin, a team of rivals. Rivalrous teams are central in modern political strategy. In February 2009, with Zimbabwe's economy in free fall, President Robert Mugabe agreed to share power with Morgan Tsvangirai, a rival he'd earlier tried to assassinate. In March 2009, Chinese president Hu Jintao named two indignantly opposing faction leaders, Xi Jinping and Li Keqiang, to help craft China's economic and political future.

I propose that the brain is best understood as a team of rivals, and the rest of this chapter will explore that framework: who the parties are, how they compete, how the union is held together, and what happens when things fall apart. As we proceed, remember that competing factions typically have the same goal—success for the country—but they often have different ways of going about it. As Lincoln put it, rivals should be turned into allies "for the sake of the greater good," and for neural subpopulations the common interest is the thriving and survival of the organism. In the same way that liberals and conservatives both love their country but can have acrimoniously different strategies for steering it, so too does the brain have competing factions that all believe they know the right way to solve problems.

THE DOMINANT TWO-PARTY SYSTEM: REASON AND EMOTION

When trying to understand the strange details of human behavior, psychologists and economists sometimes appeal to a "dual-process" account.[9] In this view, the brain contains two separate systems: one is fast, automatic, and below the surface of conscious awareness, while the other is slow, cognitive, and conscious. The first system can be labeled automatic, implicit, heuristic, intuitive, holistic, reactive, and impulsive, while the second system is cognitive, systematic, explicit, analytic, rule-based, and reflective.[10] These two processes are always battling it out.

Despite the "dual-process" moniker, there is no real reason to assume that there are only two systems—in fact, there may be several systems. For example, in 1920 Sigmund Freud suggested three competing parts in his model of the psyche: the id (instinctive), the ego (realistic and organized), and the superego (critical and moralizing).[11] In the 1950s, the American neuroscientist Paul MacLean suggested that the brain is made of three layers representing successive stages of evolutionary development: the reptilian brain (involved in survival behaviors), the limbic system (involved in emotions), and the neocortex (used in higher-order thinking). The details of both of these theories have largely fallen out of favor among neuroanatomists, but the heart of the idea survives: brains are made of competing subsystems. We will proceed using the generalized dual-process model as a starting point, because it adequately conveys the thrust of the argument.

Although psychologists and economists think of the different systems in abstract terms, modern neuroscience strives for an anatomical grounding. And it happens that the wiring diagram of the brain lends itself to divisions that generally map onto the dual-process model.[12] Some areas of your brain are involved in higher-order operations regarding events in the outside world (these include, for example, the surface of the brain just inside your temples, called the dorsolateral prefrontal cortex). In contrast, other areas are involved with monitoring your internal state, such as your level of hunger, sense of motivation, or whether something is rewarding to you (these areas include, for example, a region just behind your forehead called the medial prefrontal cortex, and several areas deep below the surface of the cortex). The situation is more complicated than this rough division would imply, because brains can simulate future states, reminisce about the past, figure out where to find things not immediately present, and so on. But for the moment, this division into systems that monitor the external and internal will serve as a rough guide, and a little later we will refine the picture.

In the effort to use labels tied neither to black boxes nor to neuroanatomy, I've chosen two that will be familiar to everyone: the *rational* and *emotional* systems. These terms are underspecified and imperfect, but they will nonetheless carry the central point about rivalries in the brain.[13] The rational system is the one that cares about analysis of things in the outside world, while the emotional system monitors internal state and worries whether things will be good or bad. In other words, as a rough guide, rational cognition involves external events, while emotion involves your internal state. You can do a math problem without consulting your internal state, but you can't order a dessert off a menu or prioritize what you feel like doing next.[14] The emotional networks are absolutely required to rank your possible next actions in the world: if you were an emotionless robot who rolled into a room, you might be able to make analyses about the objects around you, but you would be frozen with indecision about what to do next. Choices about the priority of actions are determined by our internal states: whether you head straight to the refrigerator, bathroom, or bedroom upon returning home depends not on the external stimuli in your home (those have not changed), but instead on your body's internal states.

A TIME FOR MATH, A TIME TO KILL

The battle between the rational and emotional systems is brought to light by what philosophers call the trolley dilemma. Consider this scenario: A trolley is barreling down the train tracks, out of control. Five workers are making repairs way down the track, and you, a bystander, quickly realize that they will all be killed by the trolley. But you also notice that there is a switch nearby that you can throw, and that will divert the trolley down a different track, where only a single worker will be killed. What do you do? (Assume there are no trick solutions or hidden information.)

If you are like most people, you will have no hesitation about throwing the switch: it's far better to have one person killed than five, right? Good choice.

Now here's an interesting twist to the dilemma: imagine that the same trolley is barreling down the tracks, and the same five workers are in harm's way—but this time you are a bystander on a footbridge that goes over the tracks. You notice that there is an obese man standing on the footbridge, and you realize that if you were to push him off the bridge, his bulk would be sufficient to stop the train and save the five workers. Do you push him off?

If you're like most people, you bristle at this suggestion of murdering an innocent person. But wait a minute. What differentiates this from your previous choice? Aren't you trading one life for five lives? Doesn't the math work out the same way?

What exactly is the difference in these two cases? Philosophers working in the tradition of Immanuel Kant have proposed that the difference lies in how people are being used. In the first scenario, you are simply reducing a bad situation (the deaths of five people) to a less bad situation (the death of one). In the case of the man on the bridge, he is being exploited as a means to an end. This is a popular explanation in the philosophy literature. Interestingly, there may be a more brain-based approach to understand the reversal in people's choices.

In the alternative interpretation, suggested by the neuroscientists Joshua Greene and Jonathan Cohen, the difference in the two scenarios pivots on the emotional component of actually touching someone—that is, interacting with him at a close distance.[15] If the problem is constructed so that the man on the footbridge can be dropped, with the flip of switch, through a trapdoor, many people will vote to let him drop. Something about interacting with the person up close stops most people from pushing the man to his death. Why? Because that sort of personal interaction activates the emotional networks. It changes the problem from an abstract, impersonal math problem into a personal, emotional decision.

When people consider the trolley problem, here's what brain imaging reveals: In the footbridge scenario, areas involved in motor planning and emotion become active. In contrast, in the track-switch scenario, only lateral areas involved in rational thinking become active. People register emotionally when they have to push someone; when they only have to tip a lever, their brain behaves like *Star Trek*'s Mr. Spock.

* * *

The battle between emotional and rational networks in the brain is nicely illustrated by an old episode of *The Twilight Zone*. I am paraphrasing from memory, but the plot goes something like this: A stranger in an overcoat shows up at a man's door and proposes a deal. "Here is a box with a single button on it. All you have to do is press the button and I will pay you a thousand dollars."

"What happens when I press the button?" the man asks.

The stranger tells him, "When you press the button, someone far away, someone you don't even know, will die."

The man suffers over the moral dilemma through the night. The button box rests on his kitchen table. He stares at it. He paces around it. Sweat clings to his brow.

Finally, after an assessment of his desperate financial situation, he lunges to the box and punches the button. Nothing happens. It is quiet and anticlimactic.

Then there is a knock at the door. The stranger in the overcoat is there, and he hands the man the money and takes the box. "Wait," the man shouts after him. "What happens now?"

The stranger says, "Now I take the box and give it to the next person. Someone far away, someone you don't even know."

The story highlights the ease of impersonally pressing a button: if the man had been asked to attack someone with his hands, he presumably would have declined the bargain.

In earlier times in our evolution, there was no real way to interact with others at a distance any farther than that allowed by hands, feet, or possibly a stick. That distance of interaction was salient and consequential, and this is what our emotional reaction reflects. In modern times, the situation differs: generals and even soldiers commonly find themselves far removed from the people they kill.

In Shakespeare's *Henry VI, Part 2*, the rebel Jack Cade challenges Lord Say, mocking the fact that he has never known the firsthand danger of the battlefield: "When struck'st thou one blow in the field?" Lord Say responds, "Great men have reaching hands: oft have I struck those that I never saw, and struck them dead." In modern times, we can launch forty Tomahawk surface-to-surface missiles from the deck of navy ships in the Persian Gulf and Red Sea with the touch of a button. The result of pushing that button may be watched by the missile operators live on CNN, minutes later, when Baghdad's buildings disappear in plumes. The proximity is lost, and so is the emotional influence. This impersonal nature of waging war makes it disconcertingly easy. In the 1960s, one political thinker suggested that the button to launch a nuclear war should be implanted in the chest of the President's closest friend. That way, should the President want to make the decision to annihilate millions of people on the other side of the globe, he would first have to physically harm his friend, ripping open his chest to get to the button. That would at least engage his emotional system in the decision making, so as to guard against letting the choice be impersonal.

Because both of the neural systems battle to control the single output channel of behavior, emotions can tip the balance of decision making. This ancient battle has turned into a directive of sorts for many people: *If it feels bad, it is probably wrong.*[16] There are many counter examples to this (for example, one may find oneself put off by another's sexual preference but still deem nothing morally wrong with that choice), but emotion nonetheless serves as a generally useful steering mechanism for decision making.

The emotional systems are evolutionarily old, and therefore shared with many other species, while the development of the rational system is more recent. But as we have seen, the novelty of the rational system does not necessarily indicate that it is, by itself, superior. Societies would *not* be better off if everyone were like Mr. Spock, all rationality and no emotion. Instead, a balance— a teaming up of the internal rivals—is optimal for brains. This is because the disgust we feel at pushing the man off the footbridge is critical to social interaction; the impassivity one feels at pressing a button to launch a Tomahawk missile is detrimental to civilization. Some balance of the emotional and rational systems is needed, and that balance may already be optimized by natural selection in human brains. To put it another way, a democracy split across the aisle may be just what you want—a takeover in either direction would almost certainly prove less optimal. The ancient Greeks had an analogy for life that captured this wisdom: you are a charioteer, and your chariot is pulled by two thunderous horses, the white horse of reason and the black horse of passion. The white horse is always trying to tug you off one side of the road, and the black horse tries to pull you off the other side. Your job is to hold on to them tightly, keeping them in check so you can continue down the middle of the road.

The emotional and rational networks battle not only over immediate moral decisions, but in another familiar situation as well: how we behave in time.

WHY THE DEVIL CAN SELL YOU FAME NOW FOR YOUR SOUL LATER

Some years ago, the psychologists Daniel Kahneman and Amos Tversky posed a deceptively simple question: If I were to offer you $100 right now or $110 a week from now, which would you choose? Most subjects chose to take $100 right then. It just didn't seem worthwhile to wait an entire week for another $10.

Then the researchers changed the question slightly: If I were to offer you $100 fifty-two weeks from now, or $110 fifty-three weeks from now, which would you choose? Here people tended to switch their preference, choosing to wait the fifty-three weeks. Note that the two scenarios are identical in that waiting one extra week earns you an extra $10. So why is there a preference reversal between the two?[17]

It's because people "discount" the future, an economic term meaning that rewards closer to now are valued more highly than rewards in the distant future. Delaying gratification is difficult. And there is something very special about *right now*—which always holds the highest value. Kahneman and Tversky's preference reversal comes about because the discounting has a particular shape: it drops off very quickly into the near future, and then flattens out a bit, as though more distant times are all about the same. That shape happens to look like the shape you would get if you combined two simpler processes: one that cares about short-term reward and one that holds concerns more distantly into the future.

That gave an idea to neuroscientists Sam McClure, Jonathan Cohen, and their colleagues. They reconsidered the preference-reversal problem in light of the framework of multiple competing systems in the brain. They asked volunteers to make these something-now-or-more-later economic decisions while in a brain scanner. The scientists searched for a system that cared about immediate gratification, and another that involved longer-term rationality. If the two operate independently, and fight against each other, that just might explain the data. And indeed, they found that some emotionally involved brain structures were highly activated by the choice of immediate or near-term rewards. These areas were associated with impulsive behavior, including drug addiction. In contrast, when participants opted for longer-term rewards with higher return, lateral areas of the cortex involved in higher cognition and deliberation were more active.[18] And the higher the activity in these lateral areas, the more the participant was willing to defer gratification.

Sometime between 2005 and 2006, the United States housing bubble burst. The problem was that 80 percent of recently issued mortgages were adjustable-rate. The subprime borrowers who had signed up for these loans suddenly found themselves stuck with higher payment rates and no way to refinance. Delinquencies soared. Between late 2007 and 2008, almost one million U.S. homes were foreclosed on. Mortgage-backed securities rapidly lost most of their value. Credit around the world tightened. The economy melted.

What did this have to do with competing systems in the brain? Subprime mortgage offers were perfectly optimized to take advantage of the I-want-it-now system: buy this beautiful house now with very low payments, impress your friends and parents, live more comfortably than you thought you could. At some point the interest rate on your adjustable-rate mortgage will go up, but that's a long way away, hidden in the mists of the future. By plugging directly into these instant-gratification circuits, the lenders were able to almost tank the American economy. As the economist Robert Shiller noted in the wake of the subprime mortgage crisis, speculative bubbles are caused by "contagious optimism, seemingly impervious to facts, that often takes hold when prices are rising. Bubbles are primarily social phenomena; until we understand and address the psychology that fuels them, they're going to keep forming."[19]

When you begin to look for examples of I-want-it-now deals, you'll see them everywhere. I recently met a man who accepted $500 while he was a college student in exchange for signing his body away to a university medical school after he dies. The students who accepted the deal all received ankle tattoos that tell the hospital, decades from now, where their bodies should be delivered. It's an easy sell for the school: $500 now feels good, while death is inconceivably distant. There is nothing wrong with donating one's body, but this serves to illustrate the archetypical dual-process conflict, the proverbial deal with the Devil: your wishes granted now for your soul in the distant future.

These sorts of neural battles often lie behind marital infidelity. Spouses make promises in a moment of heartfelt love, but later can find themselves in a situation in which present temptations tip their decision making the other way. In November 1995, Bill Clinton's brain decided that risking the future leadership of the free world was counterbalanced by the pleasure he had the opportunity to experience with the winsome Monica in the present moment.

So when we talk about a virtuous person, we do not necessarily mean someone who is not tempted but, instead, someone who is able to *resist* that temptation. We mean someone who does not let that battle tip to the side of instant gratification. We value such people because it is easy to yield to impulses, and inordinately difficult to ignore them. Sigmund Freud noted that arguments stemming from the intellect or from morality are weak when pitted against human passions and desires,[20] which is why campaigns to "just say no" or practice abstinence will never work. It has also been proposed that this imbalance of reason and emotion may explain the tenacity of religion in societies: world religions are optimized to tap into the emotional networks, and great arguments of reason amount to little against such magnetic pull. Indeed, the Soviet attempts to squelch religion were only partially successful, and no sooner had the government collapsed than the religious ceremonies sprang richly back to life.

The observation that people are made of conflicting short- and long-term desires is not a new one. Ancient Jewish writings proposed that the body is composed of two interacting parts: a body (*guf*), which always wants things now, and a soul (*nefesh*), which maintains a longer-term view. Similarly, Germans use a fanciful expression for a person trying to delay gratification: he must overcome his *innerer schweinehund*—which translates, sometimes to the puzzlement of English speakers, as "inner pigdog."

Your behavior—what you do in the world—is simply the end result of the battles. But the story gets better, because the different parties in the brain can learn about their interactions with one another. As a result, the situation quickly surpasses simple arm wrestling between short- and long-term desires and enters the realm of a surprisingly sophisticated process of negotiation.

THE PRESENT AND FUTURE ULYSSES

In 1909, Merkel Landis, treasurer of the Carlisle Trust Company in Pennsylvania, went on a long walk and was struck with a new financial idea. He would start something called a Christmas club. Customers would deposit money with the bank throughout the year, and there would be a fee if they withdrew their money early. Then, at the end of the year, people could access their money just in time for holiday shopping. If the idea worked, the bank would have plenty of capital to reinvest and profit from all year. But would it work? Would people willingly give up their capital all year for little or no interest?

Landis tried it, and the concept immediately caught fire. That year, almost four hundred patrons of the bank socked away an average of $28 each—quite a bit of money in the early 1900s. Landis and the other bankers couldn't believe their luck. Patrons *wanted* them to hold on to their money.

The popularity of Christmas banking clubs grew quickly, and banks soon found themselves battling each other for the holiday nest egg business. Newspapers exhorted parents to enroll their children in Christmas clubs "to develop self-reliance and the saving habit."[21] By the 1920s, several banks, including the Dime Saving Bank of Toledo, Ohio, and the Atlantic Country Trust Co. in Atlantic City, New Jersey, began manufacturing attractive brass Christmas club tokens to entice new customers.[22] (The Atlantic City tokens read, "Join our Christmas Club and Have Money When You Need It Most.")

But why did Christmas clubs catch on? If the depositors controlled their own money throughout the year, they could earn better interest or invest in emerging opportunities. Any economist would advise them to hold on to their own capital. So why would people willingly ask a bank to take away their money, especially in the face of restrictions and early withdrawal fees? The answer is obvious: people wanted someone to stop them from spending their money. They knew that if they held on to their own money, they were likely to blow it.[23]

For this same reason, people commonly use the Internal Revenue Service as an ersatz Christmas club: by claiming fewer deductions on their paychecks, they allow the IRS to keep more of their money during the year. Then, come next April, they receive the joy of a check in the mailbox. It feels like free money—but of course it's only your own. And the government got to earn interest on it instead of you. Nonetheless, people choose this route when they intuit that the extra money will burn a hole in their pocket during the year. They'd rather grant someone else the responsibility to protect them from impulsive decisions.

Why don't people take control of their own behavior and enjoy the opportunities of commanding their own capital? To understand the popularity of the Christmas club and IRS phenomena, we need to step back three millennia to the king of Ithaca and a hero of the Trojan War, Ulysses.

After the war, Ulysses was on a protracted sea voyage back to his home island of Ithaca when he realized he had a rare opportunity in front of him. His ship would be passing the island of Sirenum scopuli, where the beautiful Sirens sang melodies so alluring they beggared the human mind. The problem was that sailors who heard this music steered toward the tricky maidens, and their ships were dashed into the unforgiving rocks, drowning all aboard.

So Ulysses hatched a plan. He knew that when he heard the music, he would be as unable to resist as any other mortal man, so he came up with an idea to deal with his *future self*. Not the present, rational Ulysses, but the future, crazed Ulysses. He ordered his men to lash him to the mast of the ship and tie him there securely. This way he would be unable to move when the music wafted over the bow of the ship. Then he had them fill their ears with beeswax so they could not be seduced by the voices of the Sirens—or hear his crazed commands. He made it clear to them that they should not respond to his entreaties and should not release him until the ship was well past the Sirens. He surmised that he would be screaming, yelling, cursing, trying to force the men to steer toward the mellifluous women—he knew that this future Ulysses would be in no position to make good decisions. Therefore, the Ulysses of sound mind structured things in such a way as to prevent himself from doing something foolish when they passed the upcoming island. It was a deal struck between the present Ulysses and the future one.

This myth highlights the way in which minds can develop a meta-knowledge about how the short- and long-term parties interact. The amazing consequence is that minds can negotiate with different time points of themselves.[24]

So imagine the hostess pressing the chocolate cake upon you. Some parts of your brain want that glucose, while others parts care about your diet; some parts look at the short-term gain, other parts at long-term strategy. The battle tips toward your emotions and you decide to dig in. But not without a contract: you'll eat it only if you promise to go to the gym tomorrow. Who's negotiating with whom? Aren't both parties in the negotiation *you*?

Freely made decisions that bind you in the future are what philosophers call a Ulysses contract.[25] As a concrete example, one of the first steps in breaking an alcohol addiction is to ensure, during sober reflection, that there is no alcohol in the house. The temptation will simply be too great after a stressful workday or on a festive Saturday or a lonely Sunday.

People make Ulysses contracts all the time, and this explains the immediate and lasting success of Merkel Landis's Christmas club. When people handed over their capital in April, they were acting with a wary eye toward their October selves, who they knew would be tempted to blow the money on something selfish instead of deferring to their generous, gift-giving December selves.

Many arrangements have evolved to allow people to proactively bind the options of their future selves. Consider the existence of websites that help you lose weight by negotiating a business deal with your future self. Here's how it works: you pay a deposit of $100 with the promise that you will lose ten pounds. If you succeed by the promised time, you get all the money back. If you don't lose the weight by that time, the company keeps the money. These arrangements work on the honor system and could easily be cheated, but nonetheless these companies are profiting. Why? Because people understand that as they come closer to the date when they can win back their money, their emotional systems will care more and more about it. They are pitting short- and long-term systems against each other.*

Ulysses contracts often arise in the context of medical decision making. When a person in good health signs an advance medical directive to pull the plug in the event of a coma, he is binding himself in a contract with a possible future self—even though it is arguable that the two selves (in health and in sickness) are quite different.

An interesting twist on the Ulysses contract comes about when someone else steps in to make a decision for you—and binds your present self in deference to your future self. These situations arise commonly in hospitals, when a patient, having just experienced a traumatic life change, such as losing a limb or a spouse, declares that she wants to die. She may demand, for example, that her doctors stop her dialysis or give her an overdose of morphine. Such cases typically go before ethics boards, and the boards usually decide the same thing: don't let the patient die, because the future patient will eventually find a way to regain her emotional footing and reclaim happiness. The ethics board here acts simply as an advocate for the rational, long-term system, recognizing that the present context allows the intellect little voice against the emotions.[26] The board essentially decides that the neural congress is unfairly tilted at the moment, and that an intervention is needed to prevent a one-party takeover. Thank goodness that we can sometimes rely on the dispassion of someone else, just as Ulysses relied on his sailors to ignore his pleas. The rule of thumb is this: when

---

*Although this system works, it strikes me that there is a way to better match this business model to the neurobiology. The problem is that weight loss demands a sustained effort, while the approaching deadline for the loss of money is always distantly in the future until the day of reckoning is suddenly upon you. In a neurally optimized model, you would lose a little money each day until you have shed the ten pounds. Each day, the amount you'd lose would increase by fifteen percent. So every day brings the immediate emotional sting of monetary loss, and the sting constantly grows worse. When you've lost the ten pounds, then you stop losing money. This encourages a sustained diet ethic over the entire time window.

you cannot rely on your own rational systems, borrow someone else's.[27] In this case, patients borrow the rational systems of the board members. The board can more easily take responsibility for protecting the future patient, as its members do not hear the emotional Siren songs in which the patient is ensnared.

OF MANY MINDS

For the purpose of illustrating the team-of-rivals framework, I have made the oversimplification of subdividing the neuroanatomy into the rational and emotional systems. But I do not want to give the impression that these are the only competing factions. Instead, they are only the beginning of the team-of-rivals story. Everywhere we look we find overlapping systems that compete.

One of the most fascinating examples of competing systems can be seen with the two hemispheres of the brain, left and right. The hemispheres look roughly alike and are connected by a dense highway of fibers called the corpus callosum. No one would have guessed that the left and right hemispheres formed two halves of a team of rivals until the 1950s, when an unusual set of surgeries were undertaken. Neurobiologists Roger Sperry and Ronald Meyers, in some experimental surgeries, cut the corpus callosum of cats and monkeys. What happened? Not much. The animals acted normal, as though the massive band of fibers connecting the two halves was not really necessary.

As a result of this success, split-brain surgery was first performed on human epilepsy patients in 1961. For them, an operation that prevented the spread of seizures from one hemisphere to the other was the last hope. And the surgeries worked beautifully. A person who had suffered terribly with debilitating seizures could now live a normal life. Even with the two halves of his brain separated, the patient did not seem to act differently. He could remember events normally and learn new facts without trouble. He could love and laugh and dance and have fun.

But something strange was going on. If clever strategies were used to deliver information only to one hemisphere and not the other, then one hemisphere could learn something while the other would not. It was as though the person had two independent brains.[28] And the patients could do different tasks at the same time, something that normal brains cannot do. For example, with a pencil in each hand, split brain patients could simultaneously draw incompatible figures, such as a circle and a triangle.

There was more. The main motor wiring of the brain crosses sides, such that the right hemisphere controls the left hand and the left hemisphere controls the right hand. And that fact allows a remarkable demonstration. Imagine that the word *apple* is flashed to the left hemisphere, while the word *pencil* is simultaneously flashed to the right hemisphere. When a split-brain patient is asked to grab the item he just saw, the right hand will pick up the apple while the left hand will simultaneously pick up the pencil. The two halves are now living their own lives, disconnected.

Researchers came to realize, over time, that the two hemispheres have somewhat different personalities and skills—this includes their abilities to think abstractly, create stories, draw inferences, determine the source of a memory, and make good choices in a gambling game. Roger Sperry, one of the neurobiologists who pioneered the split-brain studies (and garnered a Nobel Prize for it), came to understand the brain as "two separate realms of conscious awareness; two sensing, perceiving, thinking and remembering systems." The two halves constitute a team of rivals: agents with the same goals but slightly different ways of going about it.

In 1976, the American psychologist Julian Jaynes proposed that until late in the second millennium B.C.E., humans had no introspective consciousness, and that instead their minds were essentially divided into two, with their left hemispheres following the commands from their right hemispheres.[29] These commands, in the form of auditory hallucinations, were interpreted as voices from the gods. About three thousand years ago, Jaynes suggests, this division of labor between the left and right hemispheres began to break down. As the hemispheres began to communicate more smoothly, cognitive processes such as introspection were able to develop. The origin of consciousness, he argues, resulted from the ability of the two hemispheres to sit down at the table together and work out their differences. No one yet knows whether Jaynes's theory has legs, but the proposal is too interesting to ignore.

The two hemispheres look almost identical anatomically. It's as though you come equipped with the same basic model of brain hemisphere in the two sides of your skull, both absorbing data from the world in slightly different ways. It's essentially one blueprint stamped out twice. And nothing could be better suited for a team of rivals. The fact that the two halves are doubles of the same basic plan is evidenced by a type of surgery called a hemispherectomy, in which one entire half of the brain is removed (this is done to treat intractable epilepsy caused by Rasmussen's encephalitis). Amazingly, as long as the surgery is performed on a child before he is about eight years old, the child is fine. Let me repeat that: the child, with only half his brain remaining, is fine. He can eat, read, speak, do math, make friends, play chess, love his parents, and everything else that a child with two hemispheres can do. Note that it is not possible to remove *any* half of the brain: you cannot remove the front half or the back half and expect survival. But the right and left halves reveal themselves as something like copies of each other. Take one away and you still have another, with roughly redundant function. Just like a pair of political parties. If the Republicans or Democrats disappeared, the other would still be able to run the country. The approach would be slightly different, but things would still work.

CEASELESS REINVENTION

I've begun with examples of rational systems versus emotional systems, and the two-factions-in-one-brain phenomenon unmasked by split-brain surgeries. But the rivalries in the brain are far more numerous, and far more subtle, than the broad-stroke ones I have introduced so far. The brain is full of smaller subsystems that have overlapping domains and take care of coinciding tasks.

Consider memory. Nature seems to have invented mechanisms for storing memory more than once. For instance, under normal circumstances, your memories of daily events are consolidated (that is, "cemented in") by an area of the brain called the hippocampus. But during frightening situations—such as a car accident or a robbery—another area, the amygdala, also lays down memories along an independent, secondary memory track.[30] Amygdala memories have a different quality to them: they are difficult to erase and they can pop back up in "flashbulb" fashion—as commonly described by rape victims and war veterans. In other words, there is more than one way to lay down memory. We're not talking about a memory of different events, but multiple memories of the *same* event—as though two journalists with different personalities were jotting down notes about a single unfolding story.

So we see that different factions in the brain can get involved in the same task. In the end, it is likely that there are even more than two factions involved, all writing down information and later competing to tell the story.[31] The conviction that memory is one thing is an illusion.

Here's another example of overlapping domains. Scientists have long debated how the brain detects motion. There are many theoretical ways to build motion detectors out of neurons, and the scientific literature has proposed wildly different models that involve connections between neurons, or the extended processes of neurons (called dendrites), or large populations of neurons.[32] The details aren't important here; what's important is that these different theories have kindled decades of debates among academics. Because the proposed models are too small to measure directly, researchers design clever experiments to support or contradict various theories. The interesting outcome has been that most of the experiments are inconclusive, supporting one model over another in some laboratory conditions but not in others. This has led to a growing recognition (reluctantly, for some) that there are *many* ways the visual system detects motion. Different strategies are implemented in different places in the brain. As with memory, the lesson here is that the brain has evolved multiple, redundant ways of solving problems.[33] The neural factions often agree about what is out there in the world, but not always. And this provides the perfect substrate for a neural democracy.

The point I want to emphasize is that biology rarely rests with a single solution. Instead, it tends to ceaselessly reinvent solutions. But why endlessly innovate—why not find a good solution and move on? Unlike the artificial intelligence laboratory, the laboratory of nature has no master programmer who checks off a subroutine once it is invented. Once the *stack block* program is coded and polished, human programmers move on to the next important step. I propose that this moving on is a major reason artificial intelligence has become stuck. Biology, in contrast to artificial intelligence, takes a different approach: when a biological circuit for *detect motion* has been stumbled upon, there is no master programmer to report this to, and so random mutation continues to ceaselessly invent new variations in circuitry, solving *detect motion* in unexpected and creative new ways.

This viewpoint suggests a new approach to thinking about the brain. Most of the neuroscience literature seeks *the* solution to whatever brain function is being studied. But that approach may be misguided. If a space alien landed on Earth and discovered an animal that could climb a tree (say, a monkey), it would be rash for the alien to conclude that the monkey is the only animal with these skills. If the alien keeps looking, it will quickly discover that ants, squirrels, and jaguars also climb trees. And this is how it goes with clever mechanisms in biology: when we keep looking, we find more. Biology never checks off a problem and calls it quits. It reinvents solutions continually. The end product of that approach is a highly overlapping system of solutions—the necessary condition for a team-of-rivals architecture.[34]

THE ROBUSTNESS OF A
MULTIPLE-PARTY SYSTEM

The members of a team can often disagree, but they do not have to. In fact, much of the time rivals enjoy a natural concordance. And that simple fact allows a team of rivals to be robust in the face of losing parts of the system. Let's return to the thought experiment of a disappearing political party. Imagine that all the key decision makers of a particular party were to die in an airplane crash, and let's consider this roughly analogous to brain damage. In many cases the loss of one party would expose the polarized, opposing opinions of a rival group—as in the case when the frontal lobes are damaged, allowing for bad behavior such as shoplifting or urinating in public. But there are other cases, perhaps much more common, in which the disappearance of a political party goes unnoticed, because all the other parties hold roughly the same opinion on some matter (for example, the importance of funding residential trash collection). This is the hallmark of a robust biological system: political parties can perish in a tragic accident and the society will still run, sometimes with little more than a hiccup to the system. It may be that for every strange clinical case in which brain damage leads to a bizarre change in behavior or perception, there are hundreds of cases in which parts of the brain are damaged with no detectable clinical sign.

An advantage of overlapping domains can be seen in the newly discovered phenomenon of *cognitive reserve*. Many people are found to have the neural ravages of Alzheimer's disease upon autopsy—but they never showed the symptoms while they were alive. How can this be? It turns out that these people continued to challenge their brains into old age by staying active in their careers, doing crossword puzzles, or carrying out any other activities that kept their neural populations well exercised. As a result of staying mentally vigorous, they built what neuropsychologists call cognitive reserve. It's not that cognitively fit people don't get

Alzheimer's; it's that their brains have protection against the symptoms. Even while parts of their brains degrade, they have other ways of solving problems. They are not stuck in the rut of having a single solution; instead, thanks to a lifetime of seeking out and building up redundant strategies, they have alternative solutions. When parts of the neural population degraded away, they were not even missed.

Cognitive reserve—and robustness in general—is achieved by blanketing a problem with overlapping solutions. As an analogy, consider a handyman. If he has several tools in his toolbox, then losing his hammer does not end his career. He can use his crowbar or the flat side of his pipe wrench. The handyman with only a couple of tools is in worse trouble.

The secret of redundancy allows us to understand what was previously a bizarre clinical mystery. Imagine that a patient sustains damage to a large chunk of her primary visual cortex, and an entire half of her visual field is now blind. You, the experimenter, pick up a cardboard shape, hold it up to her blind side, and ask her, "What do you see here?"

She says, "I have no idea—I'm blind in that half of my visual field."

"I know," you say. "But take a guess. Do you see a circle, square, or triangle?"

She says, "I really can't tell you. I don't see anything at all. I'm blind there."

You say, "I know, I know. But *guess*."

Finally, with exasperation, she guesses that the shape is a triangle. And she's *correct*, well above what random chance would predict.[35] Even though she's blind, she can tease out a hunch—and this indicates that *something* in her brain is seeing. It's just not the conscious part that depends on the integrity of her visual cortex. This phenomenon is called blindsight, and it teaches us that when conscious vision is lost, there are still subcortical factory workers behind the scenes running their normal programs. So removal of parts of the brain (in this case, the cortex) reveals underlying structures that do the same thing, just not as well. And from a neuroanatomical point of view, this is not surprising: after all, reptiles can see even though they have no cortex at all. They don't see as well as we do, but they see.[36]

*   *   *

Let's pause for a moment to consider how the team-of-rivals framework offers a different way of thinking about the brain than is traditionally taught. Many people tend to assume that the brain will be divisible into neatly labeled regions that encode, say, faces, houses, colors, bodies, tool use, religious fervor, and so on. This was the hope of the early-nineteenth-century science of phrenology, in which bumps on the skull were assumed to represent something about the size of the underlying areas. The idea was that each spot in the brain could be assigned a label on the map.

But biology rarely, if ever, pans out that way. The team-of-rivals framework presents a model of a brain that possesses multiple ways of representing the same stimulus. This view rings the death knell for the early hopes that each part of the brain serves an easily labeled function.

Note that the phrenological impulse has crept back into the picture because of our newfound power to visualize the brain with neuroimaging. Both scientists and laypeople can find themselves seduced into the easy trap of wanting to assign each function of the brain to a specific location. Perhaps because of pressure for simple sound bites, a steady stream of reports in the media (and even in the scientific literature) has created the false impression that the brain area for such-and-such has just been discovered. Such reports feed popular expectation and hope for easy labeling, but the true situation is much more interesting: the continuous networks of neural circuitry accomplish their functions using multiple, independently discovered strategies. The brain lends itself well to the complexity of the world, but poorly to clear-cut cartography.

KEEPING THE UNION TOGETHER: CIVIL WARS IN THE BRAIN DEMOCRACY

In the campy cult movie *Evil Dead 2*, the protagonist's right hand takes on a mind of its own and tries to kill him. The scene degenerates into a rendition of what you might find on a sixth-grade playground: the hero uses his left hand to hold back his right hand, which is trying to attack his face. Eventually he cuts off the hand with a chain saw and traps the still-moving hand under an upside-down garbage can. He stacks books on top of the can to pin it down, and the careful observer can see that the topmost book is Hemingway's *A Farewell to Arms*.

As preposterous as this plotline may seem, there is, in fact, a disorder called *alien hand syndrome*. While it's not as dramatic as the *Evil Dead* version, the idea is roughly the same. In alien hand syndrome, which can result from the split-brain surgeries we discussed a few pages ago, the two hands express conflicting desires. A patient's "alien" hand might pick up a cookie to put it in his mouth, while the normally behaving hand will grab it at the wrist to stop it. A struggle ensues. Or one hand will pick up a newspaper, and the other will slap it back down. Or one hand will zip up a jacket, and the other will unzip it. Some patients with alien hand syndrome have found that yelling "Stop!" will cause the other hemisphere (and the alien hand) to back down. But besides that little modicum of control, the hand is running on its own inaccessible programs, and that is why it's branded as alien—because the conscious part of the patient seems to have no predictive power over it; it does not feel as though it's part of the patient's personality at all. A patient in this situation often says, "I swear I'm not doing this." Which revisits one of the main points of this book: who is the *I*? His own brain is doing it, not anyone else's. It's simply that he doesn't have conscious access to those programs.

What does alien hand syndrome tell us? It unmasks the fact that we harbor mechanical, "alien" subroutines to which we have no access and of which we have no acquaintance. Almost all of our actions—from producing speech to picking up a mug of coffee—are run by alien subroutines, also known as zombie systems. (I use these terms interchangeably: *zombie* emphasizes the lack of conscious access, while *alien* emphasizes the foreignness of the programs.)[37] Some alien subroutines are instinctual, while some are learned; all of the highly automated algorithms that we saw in Chapter 3 (serving the tennis ball, sexing the chicks) become inaccessible zombie programs when they are burned down into the circuitry. When a professional baseball player connects his bat with a pitch that is traveling too fast for his conscious mind to track, he is leveraging a well-honed alien subroutine.

Alien hand syndrome also tells us that under normal circumstances, all the automated programs are tightly controlled such that only one behavioral output can happen at a time. The alien hand highlights the normally seamless way in which the brain keeps a lid on its internal conflicts. It requires only a little structural damage to uncover what is happening beneath. In other words, keeping the union of subsystems together is not something the brain does without effort—instead, it is an active process. It is only when factions begin to secede from the union that the alienness of the parts becomes obvious.

A good illustration of conflicting routines is found in the Stroop test, a task that could hardly have simpler instructions: name the color of the *ink* in which a word is printed. Let's say I present the word JUSTICE written in blue letters. You say, "Blue." Now I show you PRINTER written in yellow. "Yellow." Couldn't be easier. But the trick comes when I present a word that is itself the name of a color. I present the word BLUE in the color green. Now the reaction is not so easy. You might blurt out, "Blue!", or you might stop yourself and sputter out, "Green!" Either way, you have a much slower reaction time—and this belies the conflict going on under the hood. This *Stroop interference* unmasks the clash between the strong, involuntary and automatic impulse to read the word and the unusual, deliberate, and effortful task demand to state the color of the print.[38]

Remember the implicit association task from Chapter 3, the one that seeks to tease out unconscious racism? It pivots on the slower-than-normal reaction time when you're asked to link something you dislike with a positive word (such as *happiness*). Just as with the Stroop task, there's an underlying conflict between deeply embedded systems.

E PLURIBUS UNUM

Not only do we run alien subroutines; we also justify them. We have ways of retrospectively telling stories about our actions as though the actions were always our idea. As an example at the beginning of the book, I mentioned that thoughts come to us and we take credit for them ("I just had a great idea!"), even though our brains have been chewing on a given problem for a long time and eventually served up the final product. We are constantly fabricating and telling stories about the alien processes running under the hood.

To bring this sort of fabrication to light, we need only look at another experiment with split-brain patients. As we saw earlier, the right and left halves are similar to each other but not identical. In humans, the left hemisphere (which contains most of the capacity to speak language) can speak about what it is feeling, whereas the mute right hemisphere can communicate its thoughts only by commanding the left hand to point, reach, or write. And this fact opens the door to an experiment regarding the retrospective fabrication of stories. In 1978, researchers Michael Gazzaniga and Joseph LeDoux flashed a picture of a chicken claw to the left hemisphere of a split-brain patient and a picture of a snowy winter scene to his right hemisphere. The patient was then asked to point at cards that represented what he had just seen. His right hand pointed to a card with a chicken, and his left hand pointed to a card with a snow shovel. The experimenters asked him why he was pointing to the shovel. Recall that his left hemisphere (the one with the capacity for language), had information only about a chicken, and nothing else. But the left hemisphere, without missing a beat, fabricated a story: "Oh, that's simple. The chicken claw goes with the chicken, and you need a shovel to clean out the chicken shed." When one part of the brain makes a choice, other parts can quickly invent a story to explain why. If you show the command "Walk" to the right hemisphere (the one without language), the patient will get up and start walking. If you stop him and ask why he's leaving, his left hemisphere, cooking up an answer, will say something like "I was going to get a drink of water."

The chicken/shovel experiment led Gazzaniga and LeDoux to conclude that the left hemisphere acts as an "interpreter," watching the actions and behaviors of the body and assigning a coherent narrative to these events. And the left hemisphere works this way even in normal, intact brains. Hidden programs drive actions, and the left hemisphere makes justifications. This idea of retrospective storytelling suggests that we come to know our own attitudes and emotions, at least partially, by inferring them from observations of our own behavior.[39] As Gazzaniga put it, "These findings all suggest that the interpretive mechanism of the left hemisphere is always hard at work, seeking the meaning of events. It is constantly looking for order and reason, even when there is none—which leads it continually to make mistakes."[40]

This fabrication is not limited to split-brain patients. Your brain, as well, interprets your body's actions and builds a story around them. Psychologists have found that if you hold a pencil between your teeth while you read something, you'll think the material is funnier; that's because the interpretation is influenced by the smile on your face. If you sit up straight instead of slouching, you'll feel happier. The brain assumes that if the mouth and spine are doing that, it must be because of cheerfulness.

*   *   *

On December 31, 1974, Supreme Court Justice William O. Douglas was debilitated by a stroke that paralyzed his left side and confined him to a wheelchair. But Justice Douglas demanded to be checked out of the hospital on the grounds that he was fine. He declared that reports of his paralysis were "a myth." When reporters expressed skepticism, he publicly invited them to join him for a hike, a move interpreted as absurd. He even claimed to be kicking football field goals with his paralyzed side. His colleagues were perplexed by Douglas' apparently delusional behavior, and they convinced him to retire from his long-held seat in the Supreme Court.

What Douglas experienced is called *anosognosia*. This term describes a total lack of awareness about an impairment, and a typical example is a patient who completely denies their very obvious paralysis. It's not that Justice Douglas was *lying*—his brain actually believed that he could move just fine. These fabrications illustrate the lengths to which the brain will go to put together a coherent narrative. When asked to place both hands on an imaginary steering wheel, a partially paralyzed and anosognosic patient will put one hand up, but not the other. When asked if both hands are on the wheel, he will say yes. When the patient is asked to clap his hands, he may move only a single hand. If asked, "Did you clap?", he'll say yes. If you point out that you didn't hear any sound and ask him to do it again, he might not do it at all; when asked why, he'll say he "doesn't feel like it." Similarly, as mentioned in Chapter 2, one can lose vision and claim to still be able to see just fine, even while being unable to navigate a room without crashing into the furniture. Excuses are made about poor balance, rearranged chairs, and so on—all the while denying the blindness. The point about anosognosia is that the patients are not lying, and are motivated neither by mischievousness nor by embarrassment; instead, their brains are fabricating explanations that provide a coherent narrative about what is going on with their damaged bodies.

But shouldn't the contradicting evidence alert these people to a problem? After all, the patient wants to move his hand, but it is not moving. He wants to clap, but he hears no sound. It turns out that alerting the system to contradictions relies critically on particular brain regions—and one in particular, called the anterior cingulate cortex. Because of these conflict-monitoring regions, incompatible ideas will result in one side or another winning out: a story will be constructed that either makes them compatible or ignores one side of the debate. In special circumstances of brain damage, this arbitration system can be damaged—and then conflict can cause no trouble to the conscious mind. This situation is illustrated by a woman I'll call Mrs. G., who had suffered quite a bit of damage to her brain tissue from a recent stroke. At the time I met her, she was recovering in the hospital with her husband by her bedside, and seemed generally in good health and spirits. My colleague Dr. Karthik Sarma had noticed the night before that when he asked her to close her eyes, she would close only one and not the other. So he and I went to examine this more carefully.

When I asked her to close her eyes, she said "Okay," and closed one eye, as in a permanent wink.

"Are your eyes closed?" I asked.

"Yes," she said.

"Both eyes?"

"Yes."

I held up three fingers. "How many fingers am I holding up, Mrs. G.?"

"Three," she said.

"And your eyes are closed?"

"Yes."

In a nonchallenging way I said, "Then how did you know how many fingers I was holding up?"

An interesting silence followed. If brain activity were audible, this is when we would have heard different regions of her brain battling it out. Political parties that wanted to believe her eyes were closed were locked in a filibuster with parties that wanted the logic to work out: *Don't you see that we can't have our eyes closed* and *be able to see out there?* Often these battles are quickly won by the party with the most reasonable position, but this does not always happen with anosognosia. The patient will say nothing and will conclude nothing—not because she is embarrassed, but because she is simply locked up on the issue. Both parties fatigue to the point of attrition, and the original issue being fought over is finally dumped. The patient will conclude nothing about the situation. It is amazing and disconcerting to witness.

I was struck with an idea. I wheeled Mrs. G. to a position just in front of the room's only mirror and asked if she could see her own face. She said yes. I then asked her to close both her eyes. Again she closed one eye and not the other.

"Are *both* your eyes closed?"

"Yes."

"Can you see yourself?"

"Yes."

Gently I said, "Does it seem possible to see yourself in the mirror if both your eyes are closed?"

Pause. *No conclusion.*

"Does it look to you like one eye is closed or that both are closed?"

Pause. *No conclusion.*

She was not distressed by the questions; nor did they change her opinion. What would have been a checkmate in a normal brain proved to be a quickly forgotten game in hers.

Cases like Mrs. G.'s allow us to appreciate the amount of work that needs to happen behind the scenes for our zombie systems to work together smoothly and come to an agreement. Keeping the union together and making a good narrative does not happen for free—the brain works around the clock to stitch together a pattern of logic to our daily lives: what just happened and what was my role in it? Fabrication of stories is one of the key businesses in which our brains engage. Brains do this with the single-minded goal of getting the multifaceted actions of the democracy to make sense. As the coin puts it, *E pluribus unum*: out of many, one.

* * *

Once you have learned how to ride a bicycle, the brain does not need to cook up a narrative about what your muscles are doing; instead, it doesn't bother the conscious CEO at all. Because everything is predictable, no story is told; you are free to think of other issues as you pedal along. The brain's storytelling powers kick into gear only when things are conflicting or difficult to understand, as for the split-brain patients or anosognosics like Justice Douglas.

In the mid-1990s my colleague Read Montague and I ran an experiment to better understand how humans make simple choices. We asked participants to choose between two cards on a computer screen, one labeled A and the other labeled B. The participants had no way of knowing which was the better choice, so they picked arbitrarily at first. Their card choice gave them a reward somewhere between a penny and a dollar. Then the cards were reset and they were asked to choose again. Picking the same card produced a different reward this time. There seemed to be a pattern to it, but it was very difficult to detect. What the participants didn't know was that the reward in each round was based on a formula that incorporated the history of their previous forty choices—far too difficult for the brain to detect and analyze.

The interesting part came when I interviewed the players afterward. I asked them what they'd done in the gambling game and why they'd done it. I was surprised to hear all types of baroque explanations, such as "The computer liked it when I switched back and forth" and "The computer was trying punish me, so I switched my game plan." In reality, the players' descriptions of their own strategies did not match what they had actually done, which turned out to be highly predictable.[41] Nor did their descriptions match the computer's behavior, which was purely formulaic. Instead, their conscious minds, unable to assign the task to a well-oiled zombie system, desperately sought a narrative. The participants weren't *lying*; they were giving the best explanation they could—just like the split-brain patients or the anosognosics.

Minds seek patterns. In a term introduced by science writer

Michael Shermer, they are driven toward "patternicity"—the attempt to find structure in meaningless data.[42] Evolution favors pattern seeking, because it allows the possibility of reducing mysteries to fast and efficient programs in the neural circuitry.

To demonstrate patternicity, researchers in Canada showed subjects a light that flashed on and off randomly and asked them to choose which of two buttons to press, and when, in order to make the blinking more regular. The subjects tried out different patterns of button pressing, and eventually the light began to blink regularly. They had succeeded! Now the researchers asked them how they'd done it. The subjects overlaid a narrative interpretation about what they'd done, but the fact is that their button pressing was wholly unrelated to the behavior of the light: the blinking would have drifted toward regularity irrespective of what they were doing.

For another example of storytelling in the face of confusing data, consider dreams, which appear to be an interpretative overlay to nighttime storms of electrical activity in the brain. A popular model in the neuroscience literature suggests that dream plots are stitched together from essentially random activity: discharges of neural populations in the midbrain. These signals tickle into existence the simulation of a scene in a shopping mall, or a glimpse of recognition of a loved one, or a feeling of falling, or a sense of epiphany. All these moments are dynamically woven into a story, and this is why after a night of random activity you wake up, roll over to your partner, and feel as though you have a bizarre plot to relate. Ever since I was a child, I have been consistently amazed at how characters in my dreams possess such specific and peculiar details, how they come up with such rapid answers to my questions, how they produce such surprising dialogue and such inventive suggestions—all manner of things I would not have invented "myself." Many times I've heard a new joke in a dream, and this impressed me greatly. Not because the joke was so funny in the sober light of day (it wasn't) but because the joke was not one I could believe that *I* would have thought of. But, at least presumably, it was my brain and no one else's cooking up these interesting plotlines.[43] Like the split-brain patients or Justice Douglas, dreams illustrate our skills at spinning a single narrative from a collection of random threads. Your brain is remarkably good at maintaining the glue of the union, even in the face of thoroughly inconsistent data.

WHY DO WE HAVE CONSCIOUSNESS AT ALL?

Most neuroscientists study animal models of behavior: how a sea slug withdraws from a touch, how a mouse responds to rewards, how an owl localizes sounds in the dark. As these circuits are scientifically brought to light, they all reveal themselves to be nothing but zombie systems: blueprints of circuitry that respond to particular inputs with appropriate outputs. If our brains were composed *only* of these patterns of circuits, why would it feel like anything to be alive and conscious? Why wouldn't it feel like nothing—like a zombie?

A decade ago, neuroscientists Francis Crick and Christof Koch asked, "Why does not our brain consist simply of a series of specialized zombie systems?"[44] In other words, why are we conscious of anything at all? Why aren't we simply a vast collection of these automated, burned-down routines that solve problems?

Crick and Koch's answer, like mine in the previous chapters, is that consciousness exists to control—and to distribute control over—the automated alien systems. A system of automated subroutines that reaches a certain level of complexity (and human brains certainly qualify) requires a high-level mechanism to allow the parts to communicate, dispense resources, and allocate control. As we saw earlier with the tennis player trying to learn how to serve, consciousness is the CEO of the company: he sets the higher-level directions and assigns new tasks. We have learned in this chapter that he doesn't need to understand the software that each department in the organization uses; nor does he need to see their detailed logbooks and sales receipts. He merely needs to know whom to call on when.

As long as the zombie subroutines are running smoothly, the CEO can sleep. It is only when something goes wrong (say, all the departments suddenly find that their business models have catastrophically failed) that the CEO is rung up. Think about *when* your conscious awareness comes online: in those situations where events in the world *violate your expectations*. When everything is going according to the needs and skills of your zombie systems, you are not consciously aware of most of what's in front of you; when suddenly they cannot handle the task, you become consciously aware of the problem. The CEO scrambles around, looking for fast solutions, dialing up everyone to find who can address the problem best.

The scientist Jeff Hawkins offers a nice example of this: after he entered his home one day, he realized that he had experienced no conscious awareness of reaching for, grasping, and turning the doorknob. It was a completely robotic, unconscious action on his part—and this was because everything about the experience (the doorknob's feel and location, the door's size and weight, and so on) was already burned down into unconscious circuitry in his brain. It was expected, and therefore required no conscious participation. But he realized that if someone were to sneak over to his house, drill the doorknob out, and replace it three inches to the right, he would notice immediately. Instead of his zombie systems getting him directly into his house with no alerts or concerns, suddenly there would be a violation of expectations—and consciousness would come online. The CEO would rouse, turn on the alarms, and try to figure out what might have happened and what should be done next.

If you think you're consciously aware of most of what surrounds you, think again. The first time you make the drive to your new workplace, you attend to everything along the way. The drive seems to take a long time. By the time you've made the drive many times, you can get yourself there without much in the way of conscious deliberation. You are now free to think about other things; you feel as though you've left home and arrived at work in the blink of an eye. Your zombie systems are experts at taking care of business as usual. It is only when you see a squirrel in the road, or a missing stop sign, or an overturned vehicle on the shoulder that you become consciously aware of your surroundings.

All of this is consistent with a finding we learned two chapters ago: when people play a new video game for the first time, their brains are alive with activity. They are burning energy like crazy. As they get better at the game, less and less brain activity is involved. They have become more energy efficient. If you measure someone's brain and see very little activity during a task, it does not necessarily indicate that they're not trying—it more likely signifies that they have worked hard in the past to burn the programs into the circuitry. Consciousness is called in during the first phase of learning and is excluded from the game playing after it is deep in the system. Playing a simple video game becomes as unconscious a process as driving a car, producing speech, or performing the complex finger movements required for tying a shoelace. These become hidden subroutines, written in an undeciphered programming language of proteins and neurochemicals, and there they lurk—for decades sometimes—until they are next called upon.

From an evolutionary point of view, the purpose of consciousness seems to be this: an animal composed of a giant collection of zombie systems would be energy efficient but *cognitively inflexible*. It would have economical programs for doing particular, simple tasks, but it wouldn't have rapid ways of switching between programs or setting goals to become expert in novel and unexpected tasks. In the animal kingdom, most animals do certain things very well (say, prying seeds from the inside of a pine cone), while only a few species (such as humans) have the flexibility to dynamically develop new software.

Although the ability to be flexible sounds better, it does not come for free—the trade-off is a burden of lengthy childrearing. To be flexible like an adult human requires years of helplessness as an infant. Human mothers typically bear only one child at a time and have to provide a period of care that is unheard-of (and impracticable) in the rest of the animal kingdom. In contrast, animals that run only a few very simple subroutines (such as "Eat foodlike things and shrink away from looming objects") adopt a different rearing strategy, usually something like "Lay lots of eggs and hope for the best." Without the ability to write new programs, their only available mantra is: If you can't outthink your opponents, outnumber them.

So are other animals conscious? Science currently has no meaningful way to make a measurement to answer that question—but I offer two intuitions. First, consciousness is probably not an all-or-nothing quality, but comes in degrees. Second, I suggest that an animal's *degree of consciousness* will parallel its intellectual flexibility. The more subroutines an animal possesses, the more it will require a CEO to lead the organization. The CEO keeps the subroutines unified; it is the warden of the zombies. To put this another way, a small corporation does not require a CEO who earns three million dollars a year, but a large corporation does. The only difference is the number of workers the CEO has to keep track of, allocate among, and set goals for.*

If you put a red egg in the nest of a herring gull, it goes berserk. The color red triggers aggression in the bird, while the shape of the egg triggers brooding behavior—as a result, it tries to simultaneously attack the egg and incubate it.[45] It's running two programs at once, with an unproductive end result. The red egg sets off sovereign and conflicting programs, wired into the gull's brain like competing fiefdoms. The rivalry is there, but the bird

---

*There may be other advantages to having a large collection of alien systems with flexible allocation. For example, it may reduce our predictability to predators. If you had only one subroutine and ran it every time, a predator would know exactly how to pick you off (think of the crocodiles grazing on the wildebeest that swim across African rivers the same way, at the same time, every year). More complex collections of alien systems enjoy not only flexibility but a better shot at unpredictability.

has no capacity to arbitrate in the service of smooth coopera-tion. Similarly, if a female stickleback trespasses onto a male's territory, the male will display attack behavior and courtship behavior simultaneously, which is no way to win over a lady. The poor male stickleback appears to be simply a bundled collection of zombie programs triggered by simple lock-and-key inputs (*Trespass! Female!*), and the subroutines have not found any method of arbitration between them. This seems to me to suggest that the herring gull and the stickleback are not particularly conscious.

I propose that a useful index of consciousness is the capacity to successfully mediate conflicting zombie systems. The more an animal looks like a jumble of hardwired input–output subroutines, the less it gives evidence of consciousness; the more it can coordinate, delay gratification, and learn new programs, the more conscious it may be. If this view is correct, in the future a battery of tests might be able to yield a rough measure of a species' degree of consciousness. Think back to the befuddled rat we met near the beginning of the chapter, who, trapped between the drive to go for the food and the impulse to run from the shock, became stuck in between and oscillated back and forth. We all know what it's like to have moments of indecision, but our human arbitration between the programs allows us to escape these conundrums and make a deci-sion. We quickly find ways of cajoling or castigating ourselves toward one outcome or the other. Our CEO is sophisticated enough to get us out of the simple lockups that can thoroughly hamstring the poor rat. This may be the way in which our conscious minds—which play only a small part in our total neural function—really shine.

THE MULTITUDES

Let's circle back to how this allows us to think about our brains in a new way—that is, how the team-of-rivals framework allows us to address mysteries that would be inexplicable if we took the point of view of traditional computer programs or artificial intelligence.

Consider the concept of a secret. The main thing known about secrets is that keeping them is unhealthy for the brain.[46] Psychologist James Pennebaker and his colleagues studied what happened when rape and incest victims, acting out of shame or guilt, chose to hold secrets inside. After years of study, Pennebaker concluded that "the act of *not* discussing or confiding the event with another may be more damaging than having experienced the event per se."[47] He and his team discovered that when subjects confessed or wrote about their deeply held secrets, their health improved, their number of doctor visits went down, and there were measurable decreases in their stress hormone levels.[48]

The results are clear enough, but some years ago I began to ask myself how to understand these findings from the point of view of brain science. And that led to a question that I realized was unaddressed in the scientific literature: what *is* a secret, neurobiologically? Imagine constructing an artificial neural network of millions of interconnected neurons—what would a secret look like here? Could a toaster, with its interconnected parts, harbor a secret? We have useful scientific frameworks for understanding Parkinson's disease, color perception, and temperature sensation—but none for understanding what it means for the brain to have and to hold a secret.

Within the team-of-rivals framework, a secret is easily understood: it is the result of struggle between competing parties in the brain. One part of the brain wants to reveal something, and another part does not want to. When there are competing votes in the brain—one for telling, and one for withholding—that defines a secret. If no party cares to tell, that's merely a boring fact; if both parties want to tell, that's just a good story. Without the framework of rivalry, we would have no way to understand a secret.*

---

*Some people are constitutionally incapable of keeping a secret, and this balance may tell us something about the battles going on inside them and which way they tip. Good spies and secret agents are those people whose battle always tips toward long-term decision making rather than the thrill of telling.

The reason a secret is experienced consciously is because it results from a rivalry. It is not business as usual, and therefore the CEO is called upon to deal with it.

The main reason not to reveal a secret is aversion to the long-term consequences. A friend might think ill of you, or a lover might be hurt, or a community might ostracize you. This concern about the outcome is evidenced by the fact that people are more likely to tell their secrets to total strangers; with someone you don't know, the neural conflict can be dissipated with none of the costs. This is why strangers can be so forthcoming on airplanes, telling all the details of their marital troubles, and why confessional booths have remained a staple in one of the world's largest religions. It may similarly explain the appeal of prayer, especially in those religions that have very personal gods, deities who lend their ears with undivided attention and infinite love.

The newest twist on this ancient need to tell secrets to a stranger can be found in the form of websites like postsecret.com, where people go to anonymously disclose their confessions. Here are some examples: "When my only daughter was stillborn, I not only thought about kidnapping a baby, I planned it out in my head. I even found myself watching new mothers with their babies trying to pick the perfect one"; "I am almost certain that your son has autism but I have no idea how to tell you"; "Sometimes I wonder why my dad molested my sister but not me. Was I not good enough?"

As you have doubtless noticed, venting a secret is usually done for its own sake, not as an invitation for advice. If the listener spots an obvious solution to some problem revealed by the secret and makes the mistake of suggesting it, this will frustrate the teller—all she *really* wanted was to tell. The act of telling a secret can itself be the solution. An open question is why the receiver of the secrets has to be human—or human-like, in the case of deities. Telling a wall, a lizard, or a goat your secrets is much less satisfying.

WHERE IS C3PO?

When I was a child, I assumed that we would have robots by now—robots that would bring us food and clean our clothes and converse with us. But something went wrong with the field of artificial intelligence, and as a result the only robot in my home is a moderately dim-witted self-directing vacuum cleaner.

Why did artificial intelligence become stuck? The answer is clear: intelligence has proven itself a tremendously hard problem. Nature has had an opportunity to try out trillions of experiments over billions of years. Humans have been scratching at the problem only for decades. For most of that time, our approach has been to cook up intelligence from scratch—but just recently the field has taken a turn. To make meaningful progress in building thinking robots, it is now clear that we need to decipher the tricks nature has figured out.

I suggest that the team-of-rivals framework will play an important role in dislodging the jammed field of artificial intelligence. Previous approaches have made the useful step of dividing labor—but the resulting programs are impotent without differing opinions. If we hope to invent robots that think, our challenge is not simply to devise a subagent to cleverly solve each problem but instead to ceaselessly reinvent subagents, each with overlapping solutions, and then to pit them against one another. Overlapping factions offer protection against degradation (think of cognitive reserve) as well as clever problem solving by unexpected approaches.

Human programmers approach a problem by assuming there's a *best* way to solve it, or that there's a way it *should* be solved by the robot. But the main lesson we can extract from biology is that it's better to cultivate a team of populations that attack the problem in different, overlapping manners. The team-of-rivals framework suggests that the best approach is to abandon the question "What's the most clever way to solve that problem?" in favor of "Are there multiple, overlapping ways to solve that problem?"

Probably the best way to cultivate a team is with an evolutionary approach, randomly generating little programs and allowing them to reproduce with small mutations. This strategy allows us to continuously discover solutions rather than trying to think up a single perfect solution from scratch. As the biologist Leslie Orgel's second law states: "Evolution is smarter than you are." If I had a law of biology, it would be: "Evolve solutions; when you find a good one, *don't stop*."

Technology has so far not taken advantage of the idea of a democratic architecture—that is, the team-of-rivals framework. Although your computer is built of thousands of specialized parts, they never collaborate or argue. I suggest that conflict-based, democratic organization—summarized as the team-of-rivals architecture—will usher in a fruitful new age of biologically inspired machinery.[49]

* * *

The main lesson of this chapter is that you are made up of an entire parliament of pieces and parts and subsystems. Beyond a collection of local expert systems, we are collections of overlapping, ceaselessly reinvented mechanisms, a group of competing factions. The conscious mind fabricates stories to explain the sometimes inexplicable dynamics of the subsystems inside the brain. It can be disquieting to consider the extent to which all of our actions are driven by hardwired systems, doing what they do best, while we overlay stories about our choices.

Note that the population of the mental society does not always vote exactly the same way each time. This recognition is often missing from discussions of consciousness, which typically assume that what it is like to be you is the same from day to day, moment to moment. Sometimes you're able to read well; other times you drift. Sometimes you can find all the right words; other times your tongue is tangled. Some days you're a stick in the mud; other days you throw caution to the wind. So who's the real you? As the French essayist Michel de Montaigne put it, "There is as much difference between us and ourselves as there is between us and others."

A nation is at any moment most readily defined by its political parties in power. But it is also defined by the political opinions it harbors in its streets and living rooms. A comprehensive understanding of a nation must include those parties that are not in power but that could rise in the right circumstances. In this same way, you are composed of your multitudes, even though at any given time your conscious headline may involve only a subset of all the political parties.

Returning to Mel Gibson and his drunken tirade, we can ask whether there is such a thing as "true" colors. We have seen that behavior is the outcome of the battle among internal systems. To be clear, I'm not defending Gibson's despicable behavior, but I am saying that a team-of-rivals brain can naturally harbor both racist and nonracist feelings. Alcohol is not a truth serum. Instead, it tends to tip the battle toward the short-term, unreflective faction— which has no more or less claim than any other faction to be the "true" one. Now, we may *care* about the unreflective faction in someone, because it defines the degree to which they're *capable* of antisocial or dangerous behavior. It is certainly rational to worry about this aspect of a person, and it makes sense to say, "Gibson is capable of anti-Semitism." In the end, we can reasonably speak of someone's "most dangerous" colors, but "true" colors may be a subtly dangerous misnomer.

With this in mind, we can now return to an accidental oversight in Gibson's apology: "There is no excuse, nor should there be any tolerance, for anyone who thinks or expresses any kind of anti-Semitic remark." Do you see the error here? Anyone who *thinks* it? I would love it if no one ever thought an anti-Semitic remark, but for better or worse we have little hope of controlling the pathologies of xenophobia that sometimes infect the alien systems. Most of what we call thinking happens well under the surface of cognitive control. This analysis is not meant to exculpate Mel

Gibson for his rotten behavior, but it *is* meant to spotlight a question raised by everything we've learned so far: if the conscious you has less control over the mental machinery than we previously intuited, what does all this mean for responsibility? It is to this question that we turn now.

6

Why Blameworthiness
Is the Wrong Question

THE QUESTIONS RAISED BY THE
MAN ON THE TOWER

On the steamy first day of August 1966, Charles Whitman took
an elevator to the top floor of the University of Texas Tower in
Austin.[1] The twenty-five-year-old climbed three flights of stairs
to the observation deck, lugging with him a trunk full of guns
and ammunition. At the top he killed a receptionist with the
butt of his rifle. He then shot at two families of tourists coming
up the stairwell before beginning to fire indiscriminately from
the deck at people below. The first woman he shot was preg-
nant. As others ran to help her, he shot them as well. He shot
pedestrians in the street and the ambulance drivers that came to
rescue them.

The night before Whitman had sat at his typewriter and
composed a suicide note:

> I do not really understand myself these days. I am supposed to
> be an average reasonable and intelligent young man. However,
> lately (I cannot recall when it started) I have been a victim of
> many unusual and irrational thoughts.

As news of the shooting spread, all Austin police officers were
ordered to the campus. After several hours, three officers and a

quickly deputized citizen worked their way up the stairs and managed to kill Whitman on the deck. Not including Whitman, thirteen people were killed and thirty-three wounded.

The story of Whitman's rampage dominated national headlines the next day. And when police went to investigate his home for clues, the story became even more grim: in the early hours of the morning before the shooting, he had murdered his mother and stabbed his wife to death in her sleep. After these first killings, he had returned to his suicide note, now writing by hand.

> It was after much thought that I decided to kill my wife, Kathy, tonight. . . . I love her dearly, and she has been a fine wife to me as any man could ever hope to have. I cannot rationally pinpoint any specific reason for doing this. . . .

Along with the shock of the murders lay another, more hidden surprise: the juxtaposition of his aberrant actions and his unremarkable personal life. Whitman was a former Eagle Scout and marine, worked as a teller in a bank, and volunteered as a scoutmaster for Austin Scout Troop 5. As a child, he'd scored 138 on the Stanford Binet IQ test, placing him in the 99th percentile. So after he launched his bloody, indiscriminate shooting from the University of Texas Tower, everyone wanted answers.

For that matter, so did Whitman. He requested in his suicide note that an autopsy be performed to determine if something had changed in his brain—because he suspected it had. A few months before the shooting, Whitman had written in his diary:

> I talked to a doctor once for about two hours and tried to convey to him my fears that I felt overcome by overwhelming violent impulses. After one session I never saw the Doctor again, and since then I have been fighting my mental turmoil alone, and seemingly to no avail.

Whitman's body was taken to the morgue, his skull was put under the bone saw, and the medical examiner lifted the brain from its vault. He discovered that Whitman's brain harbored a tumor about the diameter of a nickel. This tumor, called a glioblastoma, had blossomed from beneath a structure called the thalamus, impinged on the hypothalamus, and compressed a third region, called the amygdala.[2] The amygdala is involved in emotional regulation, especially as regards fear and aggression. By the late 1800s, researchers had discovered that damage to the amygdala caused emotional and social disturbances.[3] In the 1930s, biologists Heinrich Klüver and Paul Bucy demonstrated that damage to the amygdala in monkeys led to a constellation of symptoms including lack of fear, blunting of emotion, and overreaction.[4] Female monkeys with amygdala damage showed inappropriate maternal behavior, often neglecting or physically abusing their infants.[5] In normal humans, activity in the amygdala increases when people are shown threatening faces, are put into frightening situations, or experience social phobias.

Whitman's intuition about himself—that something in his brain was changing his behavior—was spot-on.

> I imagine it appears that I brutally killed both of my loved ones. I was only trying to do a quick thorough job. . . . If my life insurance policy is valid please pay off my debts . . . donate the rest anonymously to a mental health foundation. Maybe research can prevent further tragedies of this type.

Others had noticed the changes as well. Elaine Fuess, a close friend of Whitman's, observed, "Even when he looked perfectly normal, he gave you the feeling of trying to control something in himself." Presumably, that "something" was his collection of angry, aggressive zombie programs. His cooler, rational parties were battling his reactive, violent parties, but damage from the tumor tipped the vote so it was no longer a fair fight.

Does the discovery of Whitman's brain tumor modify your

feelings about his senseless murdering? If Whitman had survived that day, would it adjust the sentencing you would consider appropriate for him? Does the tumor change the degree to which you consider it "his fault"? Couldn't you just as easily be unlucky enough to develop a tumor and lose control of your behavior?

On the other hand, wouldn't it be dangerous to conclude that people with a tumor are somehow free of guilt, or that they should be let off the hook for their crimes?

The man on the tower with the mass in his brain gets us right into the heart of the question of blameworthiness. To put it in the legal argot: was he *culpable*? To what extent is someone at fault if his brain is damaged in ways about which he has no choice? After all, we are not independent of our biology, right?

CHANGE THE BRAIN, CHANGE THE PERSON: THE UNEXPECTED PEDOPHILES, SHOPLIFTERS AND GAMBLERS

Whitman's case is not isolated. At the interface between neuroscience and law, cases involving brain damage crop up increasingly often. As we develop better technologies for probing the brain, we detect more problems.

Take the case of a 40-year-old man we'll call Alex, whose sexual preferences suddenly began to transform. He developed an interest in child pornography—and not just a little interest, but an overwhelming one. He poured his time into child-pornography Web sites and magazines. He also solicited prostitution at a massage parlor, something he said he had never previously done. He reported later that he'd wanted to stop, but "the pleasure principle overrode" his restraint. He worked to hide his acts, but subtle sexual advances toward his prepubescent stepdaughter alarmed his wife, who soon discovered his collection of child pornography. He was removed from his house, found guilty of child molestation, and sentenced to rehabilitation in lieu of prison.

In the rehabilitation program, he made inappropriate sexual advances toward the staff and other clients, and was expelled and routed toward prison.

At the same time, Alex was complaining of worsening headaches. The night before he was to report for prison sentencing, he couldn't stand the pain anymore, and took himself to the emergency room. He underwent a brain scan, which revealed a massive tumor in his orbitofrontal cortex. Neurosurgeons removed the tumor. Alex's sexual appetite returned to normal.

The lesson of Alex's story is reinforced by its unexpected follow-up. About six months after the brain surgery, his pedophilic behavior began to return. His wife took him back to the doctors. The neuro-radiologist discovered that a portion of the tumor had been missed in the surgery and was regrowing—and Alex went back under the knife. After the removal of the remaining tumor, his behavior returned to normal.

Alex's story highlights a deep central point: when your biology changes, so can your decision making, your appetites, and your desires. The drives you take for granted ("I'm a hetero/homosexual," "I'm attracted to children/adults," "I'm aggressive/not aggressive," and so on) depend on the intricate details of your neural machinery. Although acting on such drives is popularly thought to be a free choice, the most cursory examination of the evidence demonstrates the limits of that assumption.

Alex's sudden pedophilia illustrates that hidden drives and desires can lurk undetected behind the neural machinery of socialization. When the frontal lobe is compromised, people become "disinhibited," unmasking the presence of the seedier elements in the neural democracy. Would it be correct to say that Alex was "fundamentally" a pedophile, merely socialized to resist his impulses? Perhaps, but before we assign labels, consider that you probably would not want to discover the alien subroutines that lurk under your own frontal cortex.

A common example of this disinhibited behavior is seen in patients with frontotemporal dementia, a tragic disease in which the frontal and temporal lobes degenerate. With the loss of the brain tissue, patients lose the ability to control the hidden impulses. To the frustration of their loved ones, these patients unearth an endless variety of ways to violate social norms: shoplifting in front of store managers, removing their clothes in public, running stop signs, breaking out in song at inappropriate times, eating food scraps found in public trash cans, or being physically aggressive or sexually transgressive. Patients with frontotemporal dementia commonly end up in courtrooms, where their lawyers, doctors, and embarrassed adult children must explain to the judge that the violation was not the perpetrator's *fault*, exactly: much of their brains had degenerated, and there is currently no medication to stop it. Fifty-seven percent of frontotemporal dementia patients display socially violating behavior that sets them up for trouble with the law, as compared to only 27 percent of Alzheimer's patients.[7]

For another example of changes in the brain leading to changes in behavior, consider what has happened in the treatment of Parkinson's disease. In 2001, families and caretakers of Parkinson's patients began to notice something strange. When patients were given a drug called pramipexole, some of them turned into gamblers.[8] And not just casual gamblers—pathological gamblers. These were patients who had never before displayed gambling behavior, and now they were flying off to Vegas. One sixty-eight-year-old man amassed losses of over $200,000 in six months at a series of casinos. Some patients became consumed with internet poker, racking up unpayable credit card bills. Many did what they could to hide the losses from their families. For some, the new addiction reached beyond gambling to compulsive eating, alcohol consumption, and hypersexuality.

What was going on? Parkinson's involves the loss of brain cells that produce a neurotransmitter known as dopamine. Pramipexole works by impersonating dopamine. But it turns out that dopamine is a chemical doing double duty in the brain. Along with its role in motor commands, it also mediates the reward systems, guiding a person toward food, drink, mates, and other things useful for survival.

Because of dopamine's role in weighing the costs and benefits of decisions, imbalances in its levels can trigger gambling, overeating, and drug addiction—behaviors that result from a reward system gone awry.[9]

Physicians now watch out for these behavioral changes as a possible side effect of dopamine drugs like pramipexole, and a warning is clearly listed on the label. When a gambling situation crops up, families and caretakers are instructed to secure the credit cards of the patient and carefully monitor their online activities and local trips. Luckily, the effects of the drug are reversible—the physician simply lowers the dosage of the drug and the compulsive gambling goes away.

The lesson is clear: a slight change in the balance of brain chemistry can cause large changes in behavior. The behavior of the patient cannot be separated from his biology. If we like to believe that people make free choices about their behavior (as in, "I don't gamble because I'm strong-willed"), cases like Alex the pedophile, the frontotemporal shoplifters, and the gambling Parkinson's patients may encourage us to examine our views more carefully. Perhaps not everyone is equally "free" to make socially appropriate choices.

WHERE YOU'RE GOING, WHERE YOU'VE BEEN

Many of us like to believe that all adults possess the same capacity to make sound choices. It's a nice idea, but it's wrong. People's brains can be vastly different—influenced not only by genetics but by the environments in which they grew up. Many "pathogens" (both chemical and behavioral) can influence how you turn out; these include substance abuse by a mother during pregnancy, maternal stress, and low birth weight. As a child grows, neglect, physical abuse, and head injury can cause problems in mental development. Once the child is grown, substance abuse and exposure to a variety of toxins can damage the brain, modifying intelligence, aggression, and decision-making abilities.[10] The major public health movement to remove lead-based paint grew out of an understanding that even low levels of lead can cause brain damage that makes children less intelligent and, in some cases, more impulsive and aggressive. How you turn out depends on where you've been. So when it comes to thinking about blameworthiness, the first difficulty to consider is that people do not choose their own developmental path.

As we'll see, this understanding does not get criminals off the hook, but it's important to lead off this discussion with a clear understanding that people have very different starting points. It is problematic to imagine yourself in the shoes of a criminal and conclude, "Well, *I* wouldn't have done that"—because if you weren't exposed to *in utero* cocaine, lead poisoning, or physical abuse, and he was, then you and he are not directly comparable. Your brains are different; you don't fit in his shoes. Even if you would like to imagine what it's like to be him, you won't be very good at it.

Who you even have the possibility to be starts well before your childhood—it starts at conception. If you think genes don't matter for how people behave, consider this amazing fact: if you are a

Average Number of Violent Crimes Committed Annually in the United States

Offense	Carrying the genes	Not carrying the genes
Aggravated Assault	3,419,000	435,000
Homicide	14,196	1,468
Armed robbery	2,051,000	157,000
Sexual assault	442,000	10,000

carrier of a particular set of genes, your probability of committing a violent crime goes up by eight hundred and eighty-two *percent*. Here are statistics from the U.S. Department of Justice, which I've broken down into two groups: crimes committed by the population that carries this specific set of genes and by the population that does not:

In other words, if you carry these genes, you're eight times more likely to commit aggravated assault, ten times more likely to commit murder, thirteen times more likely to commit armed robbery, and forty-four times more likely to commit sexual assault.

About one-half of the human population carries these genes, while the other half does not, making the first half much more dangerous indeed. It's not even a contest. The overwhelming majority of prisoners carry these genes, as do 98.4 percent of those on death row. It seems clear enough that the carriers are strongly predisposed toward a different type of behavior—and these statistics alone indicate that we cannot presume that everyone is coming to the table equally equipped in terms of drives and behaviors.

We'll return to these genes in a moment, but first I want to tie the issue back to the main point we've seen throughout the book: we are not the ones driving the boat of our behavior, at least not nearly as much as we believe. *Who we are* runs well below the surface of our conscious access, and the details reach back in time before our birth, when the meeting of a sperm and egg granted us with certain attributes and not others. *Who we can be* begins with our molecular blueprints—a series of alien codes penned in invisibly small strings of acids—well before we have anything to do with it. We are a product of our inaccessible, microscopic history.

By the way, as regards that dangerous set of genes, you've probably heard of them. They are summarized as the Y chromosome. If you're a carrier, we call you a male.

*   *   *

When it comes to nature and nurture, the important point is that *you choose neither one.* We are each constructed from a genetic blueprint and born into a world of circumstances about which we have no choice in our most formative years. The complex interactions of genes and environment means that the citizens of our society possess different perspectives, dissimilar personalities, and varied capacities for decision making. These are not free-will *choices* of the citizens; these are the hands of cards we're dealt.

Because we did not choose the factors that affected the formation and structure of our brain, the concepts of free will and personal responsibility begin to sprout with question marks. Is it meaningful to say that Alex made bad *choices*, even though his brain tumor was not his fault? Is it justifiable to say that the patients with frontotemporal dementia or Parkinson's should be *punished* for their bad behavior?

If it seems we're heading in an uncomfortable direction—one that lets criminals off the hook—please read on, because I'm going to show the logic of a new argument piece by piece. The upshot will be that we can have an evidence-based legal system in which we will continue to take criminals off the streets, but we will change our reasons for punishment and our opportunities for rehabilitation. When modern brain science is laid out clearly, it is difficult to justify how our legal system can continue to function without it.

THE QUESTION OF FREE WILL, AND WHY THE ANSWER MAY NOT MATTER

"Man is a masterpiece of creation, if only because no amount of determinism can prevent him from believing that he acts as a free being."   —Georg C. Lichtenberg, *Aphorisms*

On August 20, 1994, in Honolulu, Hawaii, a female circus elephant named Tyke was performing in front of a crowd of hundreds. At some point, for reasons masked in elephant neurocircuitry, she snapped. She gored her groomer, Dallas Beckwith, and then trampled her trainer, Allen Beckwith. In front of the terrified crowd, Tyke burst through the barriers of the arena; once outside, she attacked a publicist named Steve Hirano. The entire series of bloody events was captured on the video cameras of the circusgoers. Tyke loped away down the streets of the Kakaako district. Over the next thirty minutes, Hawaiian police officers gave chase, firing a total of eighty-six shots at the elephant. Eventually, the damage added up and Tyke collapsed, dead.

Elephant gorings like this are not rare, and the most bizarre parts of their stories are the endings. In 1903, Topsy the elephant killed three of his handlers on Coney Island and, in a display of new technology, was electrocuted by Thomas Edison. In 1916, Mary the elephant, a performer with the Sparks World Famous Shows, killed her keeper in front of a crowd in Tennessee. Responding to the bloodthirsty demands of the community, the circus owner had Mary hung on a massive noose from a railroad derrick car, the only known elephant-hanging in history.

We do not even bother to ask the question of blame in regards to an off-kilter circus elephant. There are no lawyers who specialize in defending elephants, no drawn-out trials, no arguments for biological mitigation. We simply deal with the elephant in the most straightforward manner to maintain public safety. After all, Tyke and Topsy and Mary are understood simply to be animals, nothing but a weighty collection of elephantine zombie systems.

In contrast, when it comes to humans the legal system rests on the assumption that we *do* have free will—and we are judged based on this perceived freedom. However, given that our neural circuitry runs fundamentally the same algorithms as those of our pachyderm cousins, does this distinction between humans and animals make sense? Anatomically, our brains are made of all the same pieces and parts, with names like *cortex, hypothalamus, reticular formation, fornix, septal nucleus*, and so on. Differences in body plans and ecological niches slightly modify the connectivity patterns—but otherwise we find in our brains the same blueprints found in elephant brains. From an evolutionary point of view, the differences between mammalian brains exist only in the minute details. So where does this freedom of choice supposedly slip into the circuitry of humans?

* * *

As far as the legal system sees it, humans are *practical reasoners.* We use conscious deliberation when deciding how to act. We make our own decisions. Thus, in the legal system, a prosecutor must not merely show a guilty act, but a guilty mind as well.[11] And as long as there is nothing hindering the mind in its control of the body, it is assumed that the actor is fully responsible for his actions. This view of the practical reasoner is both intuitive and—as should be clear by this point in the book—deeply problematic. There is a tension between biology and law on this intuition. After all, we are driven to be who we are by vast and complex biological networks. We do not come to the table as blank slates, free to take in the world and come to open-ended decisions. In fact, it is not clear how much the conscious *you*—as opposed to the genetic and neural you—gets to do any deciding at all.

We've reached the crux of the issue. How exactly should we assign culpability to people for their varied behavior, when it is difficult to argue that the choice was ever really available?

Or *do* people have a choice about how they act, despite it all? Even in the face of all the machinery that constitutes you, is there some small internal voice that is independent of the biology, that directs decisions, that incessantly whispers the right thing to do? Isn't this what we call free will?

* * *

The existence of free will in human behavior is the subject of an ancient and heated debate. Those who support free will typically base their argument on direct subjective experience (I *feel* like I made the decision to lift my finger just now), which, as we are about to see, can be misleading. Although our decisions may seem like free choices, no good evidence exists that they actually are.

Consider a decision to move. It feels as though free will leads you to stick out your tongue, or scrunch up your face, or call someone a name. But free will is not *required* to play any role in these acts. Take Tourette's syndrome, in which a person suffers from involuntary movements and vocalizations. A typical Touretter may stick out his tongue, scrunch up his face, call someone a name—all without *choosing* to do so. A common symptom of Tourette's is called coprolalia, an unfortunate behavior in which the person bursts out with socially unacceptable words or phrases, such as curse words or racial epithets. Unfortunately for the Tourette's patient, the words coming out of their mouths are usually the last things they would want to say in that situation: the coprolalia is triggered by seeing someone or something that makes the exclamation forbidden. For example, upon seeing an obese person they may be compelled to shout "Fatso!" The forbidden quality of the thought drives the compulsion to shout it out.

The motor tics and inappropriate exclamations of Tourette's are not generated with what we would call free will. So we immediately learn two things from the Tourette's patient. First, sophisticated action can occur in the absence of free will. This means that witnessing a complicated act in ourselves or someone else should not convince us that there was free will behind it. Second, the Tourette's patient cannot *not* do it: they cannot use free will to override or control what other parts of their brain have decided to do. They have no *free won't*. What the lack of free will and the lack of free won't have in common is the lack of "free." Tourette's syndrome provides a case in which the zombie systems make decisions and we all agree that the person is not responsible.

Such a lack of free decisions is not restricted to Tourette's. We see this also with so-called psychogenic disorders in which movements of the hands, arms, legs, and face are involuntary, even though they certainly *look* voluntary: ask such a patient why she is moving her fingers up and down, and she will explain that she has no control over her hand. She cannot not do it. Similarly, as we saw in the previous chapter, split-brain patients can often develop alien hand syndrome: while one hand buttons up a shirt, the other hand works to unbutton it. When one hand reaches for a pencil, the other bats it away. No matter how hard the patient tries, he cannot make his alien hand *not* do what it's doing. The decisions are not "his" to freely start or stop.

Unconscious acts are not limited to unintended shouts or wayward hands; they can be surprisingly sophisticated. Consider Kenneth Parks, a twenty-three-year-old Toronto man with a wife, a five-month-old daughter, and a close relationship with his in-laws. Suffering from financial difficulties, marital problems, and a gambling addiction, he made plans to go see his in-laws to talk about his troubles. His mother-in-law, who described him as a "gentle giant," was looking forward to discussing his issues with him. But a day before their meeting, in the wee hours of the morning of May 23, 1987, Kenneth got out of bed, but did not awaken. Sleepwalking, he climbed into his car and drove fourteen miles to his in-laws' home. He broke in and stabbed his mother-in-law to death, and then assaulted his father-in-law, who survived. Afterward, he drove himself to the police station. Once there, he said, "I think I have killed some people . . . my hands," realizing for the first time that his own hands were severely cut. He was taken to the hospital, where the tendons of his hands were operated upon.

Over the next year, Kenneth's testimony was remarkably consistent even in the face of attempts to lead him astray: he remembered nothing of the incident. Moreover, while all parties agreed that Kenneth had undoubtedly committed the murder, they also agreed that he had no motive for the crime. His defense attorneys argued that this was a case of killing while sleepwalking, known as homicidal somnambulism.[12]

At the court hearing in 1988, psychiatrist Ronald Billings gave the following expert testimony:

Q. Is there any evidence that a person could formulate a plan while they were awake and then in some way ensure that they carry it out in their sleep?

A. No, absolutely not. Probably the most striking feature of what we know of what goes on in the mind during sleep is that it's very independent of waking mentation in terms of its objectives and so forth. There is a lack of control of directing our minds in sleep compared to wakefulness. In the waking state, of course, we often voluntarily plan things, what we call volition—that is, we decide to do this as opposed to that—and there is no evidence that this occurs during the sleepwalking episode. . . .

Q. And assuming he was sleepwalking at the time, would he have the capacity to intend?

A. No.

Q. Would he have appreciated what he was doing?

A. No, he would not.

Q. Would he have understood the consequences of what he was doing?

A. No, I do not believe that he would. I think it would all have been an unconscious activity, uncontrolled and unmeditated.

Homicidal sleepwalking has proven a difficult challenge for the courts, because while the public reaction is to cry "Faker!", the brain does in fact operate in a different state during sleep, and sleepwalking is a verifiable phenomenon. In disorders of sleep, known as parasomnias, the enormous networks of the brain do not always transition seamlessly between the sleeping and waking states—they can become stuck in between. Given the colossal amount of neural coordination required for the transition (including the changing patterns of neurotransmitter systems, hormones, and electrical activity), it is perhaps surprising that parasomnias are not more common than they are.

While the brain normally emerges from slow-wave sleep into lighter stages, and finally to wakefulness, Kenneth's electroencephalogram (EEG) showed a problem in which his brain tried to emerge straight from a deep sleep stage directly into wakefulness—and it attempted this hazardous transition ten to twenty times per night. In a normal sleeping brain, such a transition is not attempted even once in a night. Because there was no way for Kenneth to fake his EEG results, these findings were the clincher that convinced the jury that he indeed suffered from a sleepwalking problem—a problem severe enough to render his actions involuntary. On May 25, 1988, the jury in the Kenneth Parks case declared him not guilty of the murder of his mother-in-law and, subsequently, of the attempted murder of his father-in-law.[13]

As with Tourette's sufferers, those subject to psychogenic disorders, and the split-brain patients, Kenneth's case illustrates that high-level behaviors can happen in the absence of free will. Like your heartbeat, breathing, blinking, and swallowing, even your mental machinery can run on autopilot.

The crux of the question is whether *all* of your actions are fundamentally on autopilot or whether there is some little bit that is "free" to choose, independent of the rules of biology. This has always been the sticking point for both philosophers and scientists. As far as we can tell, all activity in the brain is driven by other activity in the brain, in a vastly complex, inter-connected network. For better or worse, this seems to leave no room for anything *other than* neural activity—that is, no room for a ghost in the machine. To consider this from the other direc-tion, if free will is to have any effect on the actions of the body, it needs to influence the ongoing brain activity. And to do that, it needs to be physically connected to at least some of the neurons. But we don't find any spot in the brain that is not itself driven by other parts of the network. Instead, every part of the brain is densely interconnected with—and driven by—other brain parts. And that suggests that no part is independent and therefore "free."

So in our current understanding of science, we can't find the physical gap in which to slip free will—the uncaused causer—

because there seems to be no part of the machinery that does not follow in a causal relationship from the other parts. Everything stated here is predicated on what we know at this moment in history, which will certainly look crude a millennium from now; however, at this point, no one can see a clear way around the problem of a nonphysical entity (free will) interacting with a physical entity (the stuff of the brain).

But let's say that you still intuit very strongly that you have free will, despite the biological concerns. Is there any way neuroscience can try to directly *test* for free will?

In the 1960s, a scientist named Benjamin Libet placed electrodes on the heads of subjects and asked them to do a very simple task: lift their finger at a time of their own choosing. They watched a high-resolution timer and were asked to note the exact moment at which they "felt the urge" to make the move.

Libet discovered that people became aware of an urge to move about a quarter of a second before they actually made the move. But that wasn't the surprising part. He examined their EEG recordings—the brain waves—and found something more surprising: the activity in their brains began to rise *before* they felt the urge to move. And not just by a little bit. By over a second. (See figure on the following page.) In other words, parts of the brain were making decisions well before the person consciously experienced the urge.[14] Returning to the newspaper analogy of consciousness, it seems that our brains crank away behind the scenes—developing neural coalitions, planning actions, voting on plans—before we receive the news that we've just had the great idea to lift a finger.

Libet's experiments caused a commotion.[15] Could it be true that the conscious mind is the last one in the chain of command to receive any information? Did his experiment drive the nail into the coffin of free will? Libet himself fretted over this possibility raised by his own experiments, and finally suggested that we might retain freedom in the form of *veto* power. In other words, while we can't control the fact that we get the urge to move our finger, perhaps we retain a tiny window of time to stop the lifting

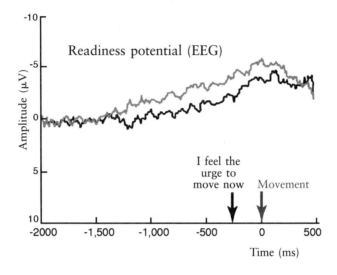

"Move your finger when the impulse grabs you." Long before a voluntary movement is enacted, a buildup of neural activity can be measured. The "readiness potential" is larger when subjects judge the time of their urge to move (grey trace), rather than the movement itself (black trace). From Eagleman, *Science*, 2004, adapted from Sirigu et al, *Nature Neuroscience*, 2004.

of our finger. Does this save free will? It's difficult to say. Despite the impression that a veto might be freely chosen, there is no evidence to suggest that it, too, wouldn't be the result of neural activity that builds up behind the scenes, hidden from conscious view.

People have proposed several other arguments to try to save the concept of free will. For example, while classical physics describes a universe that is strictly deterministic (each thing follows from the last in a predictable way), the quantum physics of the atomic scale introduces unpredictability and uncertainty as an inherent part of the cosmos. The fathers of quantum physics wondered whether this new science might save free will. Unfortunately, it doesn't. A system that is probabilistic and unpredictable is every bit as unsatisfying as a system that is deterministic, because in both cases there's no choice. It's either coin flips or billiard balls, but neither case equates to freedom in the sense that we'd desire to have it.

Other thinkers trying to save free will have looked to chaos theory, pointing out that the brain is so vastly complex that there is no way, in practice, to determine its next moves. While this is certainly true, it doesn't meaningfully address the free-will problem, because the systems studied in chaos theory are still deterministic: one step leads inevitably to the next. It is very difficult to predict where chaotic systems are going, but each state of the system is causally related to the previous state. It is important to stress the difference between a system being unpredictable and it being free. In the collapse of a pyramid of ping-pong balls, the complexity of the system makes it impossible to predict the trajectories and final positions of the balls—but each ball nonetheless follows the deterministic rules of motion. Just because we can't say where it's all going does not mean that the collection of balls is "free."

So despite all our hopes and intuitions about free will, there is currently no argument that convincingly nails down its existence.

*　　*　　*

The question of free will matters quite a bit when we turn to culpability. When a criminal stands in front of the judge's bench having recently committed a crime, the legal system wants to know whether he is *blameworthy*. After all, whether he is fundamentally responsible for his actions navigates the way we punish. You might punish your child if she writes with a crayon on the wall, but you wouldn't punish her if she did the same thing while sleepwalking. But why not? She's the same child with the same brain in both cases, isn't she? The difference lies in your intuitions about free will: in one case she has it, in the other she doesn't. In one she's choosing to act mischievously, in the other she's an unconscious automaton. You assign culpability in the first case and not in the second.

The legal system shares your intuition: responsibility for your actions parallels volitional control. If Kenneth Parks was awake when he killed his in-laws, he hangs. If asleep, he's acquitted. Similarly, if you hit someone in the face, the law cares whether you were being aggressive or if you have hemiballismus, a disorder in which your limbs can flail wildly without warning. If you crash your truck into a roadside fruit stand, the law cares whether you were driving like a maniac or instead were the victim of a heart attack. All these distinctions pivot on the assumption that we possess free will.

But do we? Don't we? Science can't yet figure out a way to say yes, but our intuition has a hard time saying no. After centuries of debate, free will remains an open, valid, and relevant scientific problem.

I propose that *the answer to the question of free will doesn't matter*—at least not for the purposes of social policy—and here's why. In the legal system, there is a defense known as an *automatism*. This is pled when the person performs an automated act— say, if an epileptic seizure causes a driver to steer into a crowd. The automatism defense is used when a lawyer claims that an act was due to a biological process over which the defendant had little or no control. In other words, there was a guilty act, but there was not a *choice* behind it.

But wait a moment. Based on what we've been learning, don't such biological processes describe most or, some would argue, all of what is going on in our brains? Given the steering power of our genetics, childhood experiences, environmental toxins, hormones, neurotransmitters, and neural circuitry, enough of our decisions are beyond our explicit control that we are arguably not the ones in charge. In other words, free will *may* exist—but if it does, it has very little room in which to operate. So I'm going to propose what I call the *principle of sufficient automatism*. The principle arises naturally from the understanding that free will, if it exists, is only a small factor riding on top of enormous automated machinery. So small that we may be able to think about bad decision making in the same way we think about any other physical process, such as diabetes or lung disease.[16] The principle states that the answer to the free-will question simply does not matter. Even if free will is conclusively proven to exist one hundred years from now, it will not change the fact that human behavior largely operates almost without regard to volition's invisible hand.

To put this another way, Charles Whitman, Alex the sudden pedophile, the frontotemporal shoplifters, the gambling Parkinson's patients, and Kenneth Parks all share the common upshot that acts cannot be considered separately from the biology of the actors. Free will is not as simple as we intuit—and our confusion about it suggests that we cannot meaningfully use it as the basis of punishment decisions.

In considering this problem, Lord Bingham, Britain's senior law lord, recently put it this way:

> In the past, the law has tended to base its approach . . . on a series of rather crude working assumptions: adults of competent mental capacity are free to choose whether they will act in one way or another; they are presumed to act rationally, and in what they conceive to be their own best interests; they are credited with such foresight of the consequences of their actions as reasonable people in their position could ordinarily be expected to have; they are generally taken to mean what they say. Whatever the merits or demerits of working assumptions such as these in the ordinary range of cases, it is evident that they do not provide a uniformly accurate guide to human behaviour.[17]

Before moving into the heart of the argument, let's put to rest the concern that biological explanations will lead to freeing criminals on the grounds that nothing is their fault. Will we still punish criminals? Yes. Exonerating all criminals is neither the future nor the goal of an improved understanding. *Explanation does not equal exculpation.* Societies will always need to get bad people off the streets. We will not abandon punishment, but we will refine the *way* we punish—as we turn to now.

THE SHIFT FROM BLAME TO BIOLOGY

The study of brains and behaviors finds itself in the middle of a conceptual shift. Historically, clinicians and lawyers have agreed on an intuitive distinction between neurological disorders ("brain problems") and psychiatric disorders ("mind problems").[18] As recently as a century ago, the prevailing attitude was to get psychiatric patients to "toughen up," either by deprivation, pleading, or torture. The same attitude applied to many disorders; for example, some hundreds of years ago, epileptics were often abhorred because their seizures were understood as demonic possessions—perhaps in direct retribution for earlier behavior.[19] Not surprisingly, this proved an unsuccessful approach. After all, while psychiatric disorders tend to be the product of more subtle forms of brain pathology, they are based, ultimately, in the biological details of the brain. The clinical community has recognized this with a shift in terminology, now referring to mental disorders under the label *organic disorders*. This term indicates that there is indeed a physical (organic) basis to the mental problem rather than a purely "psychic" one, which would mean that it has no relation to the brain—a concept that nowadays makes little sense.

What accounts for the shift from blame to biology? Perhaps the largest driving force is the effectiveness of the pharmaceutical treatments. No amount of beating will chase away depression, but a little pill called fluoxetine often does the trick. Schizophrenic symptoms cannot be overcome by exorcism, but can be controlled by risperidone. Mania responds not to talking or to ostracism, but to lithium. These successes, most of them introduced in the past sixty years, have underscored the idea that it does not make sense to call some disorders brain problems while consigning others to the ineffable realm of the psychic. Instead, mental problems have begun to be approached in the same way we might approach a broken leg. The neuroscientist Robert Sapolsky invites us to consider this conceptual shift with a series of questions:

> Is a loved one, sunk in a depression so severe that she cannot function, a case of a disease whose biochemical basis is as "real" as is the biochemistry of, say, diabetes, or is she merely indulging herself? Is a child doing poorly at school because he is unmotivated and slow, or because there is a neurobiologically based learning disability? Is a friend, edging towards a serious problem with substance abuse, displaying a simple lack of discipline, or suffering from problems with the neurochemistry of reward?[20]

The more we discover about the circuitry of the brain, the more the answers tip away from accusations of indulgence, lack of motivation, and poor discipline—and move toward the details of the biology. The shift from blame to science reflects our modern understanding that our perceptions and behaviors are controlled by inaccessible subroutines that can be easily perturbed, as seen with the split-brain patients, the frontotemporal dementia victims, and the Parkinsonian gamblers. But there's a critical point hidden in here. Just because we've shifted away from blame does not mean we have a full understanding of the biology.

Although we know that there is a strong relationship between brain and behavior, neuroimaging remains a crude technology, unable to meaningfully weigh in on assessments of guilt or innocence, especially on an individual basis. Imaging methods make use of highly processed blood-flow signals, which cover tens of cubic millimeters of brain tissue. In a single cubic millimeter of brain tissue, there are some one hundred million synaptic connections between neurons. So modern neuroimaging is like asking an astronaut in the space shuttle to look out the window and judge how America is doing. He can spot giant forest fires, or a plume of volcanic activity billowing from Mount Rainier, or the consequences of broken New Orleans levees—but from his vantage point he is unable to detect whether a crash of the stock market has led to widespread depression and suicide, whether racial tensions have sparked rioting, or whether the population has been stricken with influenza. The astronaut doesn't have the resolution to discern those details, and neither does the modern neuroscientist have the resolution to make detailed statements about the health of the brain. He can say nothing about the minutiae of the microcircuitry, nor the algorithms that run on the vast seas of millisecond-scale electrical and chemical signaling.

For example, a study by psychologists Angela Scarpa and Adrian Raine found that there are measurable differences in the brain activity of convicted murderers and control subjects, but these differences are subtle and reveal themselves only in group measurement. Therefore, they have essentially no diagnostic power for an individual. The same goes for neuroimaging studies with psychopaths: measurable differences in brain anatomy apply on a population level but are currently useless for individual diagnosis.[21]

And this puts us in a strange situation.

THE FAULT LINE: WHY BLAMEWORTHINESS IS THE WRONG QUESTION

Consider a common scenario that plays out in courtrooms around the world: A man commits a criminal act; his legal team detects no obvious neurological problem; the man is jailed or sentenced to death. But *something* is different about the man's neurobiology. The underlying cause could be a genetic mutation, a bit of brain damage caused by an undetectably small stroke or tumor, an imbalance in neurotransmitter levels, a hormonal imbalance—or any combination. Any or all of these problems may be undetectable with our current technologies. But they can cause differences in brain function that lead to abnormal behavior.

Again, an approach from the biological view point does not mean that the criminal will be exculpated; it merely underscores the idea that his actions are not divorced from the machinery of his brain, just as we saw with Charles Whitman and Kenneth Parks.

We don't blame the sudden pedophile for his tumor, just as we don't blame the frontotemporal shoplifter for the degeneration of his frontal cortex.[22] In other words, if there is a measurable brain problem, that buys leniency for the defendant. He's not really to blame.

But we *do* blame someone if we lack the technology to detect a biological problem. And this gets us to the heart of our argument: *that blameworthiness is the wrong question to ask.*

Imagine a spectrum of culpability. On one end, you have people like Alex the pedophile, or a patient with frontotemporal dementia who exposes himself to schoolchildren. In the eyes of the judge and jury, these are people who suffered brain damage at the hands of fate and did not choose their neural situation.

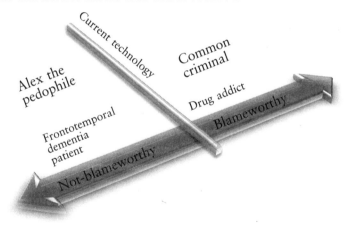

On the blameworthy side of the fault line is the common criminal, whose brain receives little study, and about whom our current technology might be able to say very little anyway. The overwhelming majority of criminals are on this side of the line, because they don't have any obvious biological problems. They are simply thought of as freely choosing actors.

Somewhere in the middle of the spectrum you might find someone like Chris Benoit, a professional wrestler whose doctor conspired with him to provide massive amounts of testosterone under the guise of hormone replacement therapy. In late June 2007, in a fit of anger known as steroid rage, Benoit came home, murdered his son and wife, and then committed suicide by hanging himself with the pulley cord of one of his weight machines. He has the biological mitigator of the hormones controlling his emotional state, but he seems more blameworthy because he chose to ingest them in the first place. Drug addicts in general are typically viewed near the middle of the spectrum: while there is some understanding that addiction is a biological issue and that drugs rewire the brain, it is also the case that addicts are often interpreted as responsible for taking the first hit.

This spectrum captures the common intuition that juries seem to have about blameworthiness. But there is a deep problem with this. Technology will continue to improve, and as we grow better at measuring problems in the brain, the fault line will drift toward the not-blameworthy-side—that is, into the territory of those we currently hold fully accountable. Problems that are now opaque will open up to examination by new techniques, and we may someday find that certain types of bad behavior will have a meaningful biological explanation—as has happened with schizophrenia, epilepsy, depression, and mania. Currently we can detect only large brain tumors, but in one hundred years we will be able to detect patterns at unimaginably small levels of the microcircuitry that correlate with behavioral problems. Neuroscience will be better able to say why people are predisposed to act the way they do. As we become more skilled at specifying how behavior results from the microscopic details of the brain, more defense lawyers will appeal to biological mitigators, and more juries will place defendants on the not-blameworthy side of the line.

A just legal system cannot define culpability simply by the limitations of current technology. A legal system that declares a person culpable at the beginning of a decade and not culpable at the end is not one in which culpability carries a clear meaning.

*   *   *

The heart of the problem is that it no longer makes sense to ask, "To what extent was it his *biology* and to what extent was it *him*?" The question no longer makes sense because we now understand those to be the same thing. There is no meaningful distinction between his biology and his decision making. They are inseparable.

As the neuroscientist Wolf Singer recently suggested: even when we cannot measure what is wrong with a criminal's brain, we can fairly safely assume that *something* is wrong.[23] His actions are *sufficient evidence* of a brain abnormality, even if we don't know (and maybe will never know) the details.[24] As Singer puts it: "As long as we can't identify all the causes, which we cannot and will probably never be able to do, we should grant that for everybody there is a neurobiological reason for being abnormal." Note that most of the time we cannot measure an abnormality in criminals. Take Eric Harris and Dylan Klebold, the shooters at Columbine High School in Colorado, or Seung-Hui Cho, the shooter at Virginia Tech. Was something wrong with their brains? We'll never know, because they—like most school shooters—were killed at the scene. But we can safely assume there was *something* abnormal in their brains. It's a rare behavior; most students don't do that.

The bottom line of the argument is that criminals should always be treated as incapable of having acted otherwise. The criminal activity itself should be taken as evidence of brain abnormality, regardless whether currently measurable problems can be pinpointed. This means that medical expert witnessing can be deeply problematic: often, such testimony reflects only whether we currently have names and measurements for problems, not whether the problem exists.

So culpability appears to be the *wrong question to ask*.

Here's the right question: What do we do, *moving forward*, with an accused criminal?

The history of a brain in front of the judge's bench can be very complex—all we ultimately want to know is how a person is likely to behave in the future.

WHAT DO WE DO FROM HERE?
A FORWARD-LOOKING, BRAIN-COMPATIBLE LEGAL SYSTEM

While our current style of punishment rests on a bedrock of personal volition and blame, the present line of argument suggests an alternative. Although societies possess deeply ingrained impulses for punishment, a forward-looking legal system would be more concerned with how to best serve the society from this day forward. Those who break the social contracts need to be warehoused, but in this case the future is of more importance than the past.[25] Prison terms do not have to be based on a desire for bloodlust, but instead can be calibrated to the risk of reoffending. Deeper biological insight into behavior will allow a better understanding of recidivism—that is, who will go out and commit more crimes. And this offers a basis for rational and evidence-based sentencing: some people need to be taken off the streets for a longer time, because their likelihood of reoffense is high; others, due to a variety of extenuating circumstances, are less likely to recidivate.

But how can we tell who presents a high risk of recidivism? After all, the details of a court trial do not always give a clear indication of the underlying troubles. A better strategy incorporates a more scientific approach.

Consider the important changes that have happened in the sentencing of sex offenders. Several years ago, researchers began to ask psychiatrists and parole board members how likely it was that individual sex offenders would relapse when let out of prison. Both the psychiatrists and the parole board members had experience with the criminals in question, as well as with hundreds before them—so predicting who was going to go straight and who would be coming back was not difficult.

Or wasn't it? The surprise outcome was that their guesses showed almost no correlation with the actual outcomes. The psychiatrists and parole board members had the predictive accuracy of coin-flipping. This result astounded the research community, especially given the expectation of well-refined intuitions among those who work directly with the offenders.

So researchers, in desperation, tried a more actuarial approach. They set about measuring dozens of factors from 22,500 sex offenders who were about to be released: whether the offender had ever been in a relationship for more than one year, had been sexually abused as a child, was addicted to drugs, showed remorse, had deviant sexual interests, and so on. The researchers then tracked the offenders for five years after release to see who ended up back in prison. At the end of the study, they computed which factors best explained the reoffense rates, and from these data they were able to build actuarial tables to be used in sentencing. Some offenders, according to the statistics, appear to be a recipe for disaster—and they are taken away from society for a longer time. Others are less likely to present a future danger to society, and they receive shorter sentences. When you compare the predictive power of the actuarial approach to that of the parole boards and psychiatrists, there is no contest: numbers win over intuitions. These actuarial tests are now used to determine the length of sentencing in courtrooms across the nation.

It will always be impossible to know with precision what someone will do upon release from prison, because real life is complicated. But more predictive power is hidden in the numbers than people customarily expect. Some perpetrators are more dangerous than others, and, despite superficial charm or superficial repugnance, dangerous people share certain patterns of behavior in common. Statistically-based sentencing has its imperfections, but it allows evidence to trump folk-intuition, and it offers sentencing customization in place of the blunt guidelines that the legal system typically employs. As we introduce brain science into these measures—for example, with neuroimaging studies—the predictive power will only improve. Scientists will never be able to foretell with high certainty who will reoffend, because that depends on multiple factors, including circumstance and opportunity. Nonetheless, good guesses are possible, and neuroscience will make those guesses better.[26]

Note that the law, even in the absence of detailed neurobiological knowledge, already embeds a bit of forward thinking: consider the lenience afforded a crime of passion versus a premeditated murder. Those who commit the former are less likely to recidivate than those who commit the latter, and their sentences sensibly reflect that.

Now, there's a critical nuance to appreciate here. Not everyone with a brain tumor undertakes a mass shooting, and not all males commit crimes. Why not? As we will see in the next chapter, it is because genes and environment interact in unimaginably complex patterns.[27] As a result, human behavior will always remain unpredictable. This irreducible complexity has consequences: when a brain is standing in front of the bench, the judge cannot care about the history of the brain. Was there fetal maldevelopment, cocaine use during pregnancy, child abuse, a high level of *in utero* testosterone, any small genetic change that offered a 2 percent higher predisposition to violence if the child was later exposed to mercury? All of these factors and hundreds of others interact, with the upshot that it would be a fruitless endeavor for the judge to try to disentangle them to determine blameworthiness. So the legal system *has* to become forward-looking, primarily because it can no longer hope to do otherwise.

*   *   *

Beyond customized sentencing, a forward-thinking legal system informed by scientific insights into the brain will enable us to stop treating prison as a one-size-fits-all solution. To be clear, I'm not opposed to incarceration, and its purpose is not limited to the removal of dangerous people from the streets. The prospect of incarceration deters many crimes, and time actually spent in prison can steer some people away from further criminal acts upon their release.

But that works only for those whose brains function normally. The problem is that prisons have become our de facto mental-health-care institutions—and inflicting punishment on the mentally ill usually has little influence on their future behavior.

An encouraging trend is the establishment of mental-health courts around the nation: through such courts, people with mental illnesses can be helped while confined in a tailored environment. Cities such as Richmond, Virginia, are moving in this direction, for reasons of justice as well as cost-effectiveness. Sheriff C. T. Woody, who estimates that nearly 20 percent of Richmond's prisoners are mentally ill, told CBS News, "The jail isn't a place for them. They should be in a mental-health facility." Similarly, many jurisdictions are opening drug courts and developing alternative sentences; they have realized that prisons are not as useful for solving addictions as are meaningful drug-rehabilitation programs.

A forward-thinking legal system will also parlay biological understanding into customized *rehabilitation*, viewing criminal behavior the way we understand other such medical conditions as epilepsy, schizophrenia, and depression—conditions that now allow the seeking and giving of help. These and other brain disorders have found themselves on the other side of the fault line now, where they rest comfortably as biological, not demonic, issues. So what about other forms of behavior, such as criminal acts? The majority of lawmakers and voters stand in favor of rehabilitating criminals instead of packing them into overcrowded prisons, but the challenge has been the dearth of new ideas about *how* to rehabilitate.

And, of course, we cannot forget the scare that still lives on in the collective consciousness: frontal lobotomies. The lobotomy (originally called a leucotomy) was invented by Egas Moniz, who thought it might make sense to help criminals by scrambling their frontal lobes with a scalpel. The simple operation cuts the connections to and from the prefrontal cortex, often resulting in major personality changes and possible mental retardation.

Moniz tested this out on several criminals and found, to his satisfaction, that it calmed them down. In fact, it flattened their personalities entirely. Moniz's protégé, Walter Freeman, noticing that institutional care was hampered by a lack of effective treatments, saw the lobotomy as an expedient tool to liberate large populations from treatment and back into private life.

Unfortunately, it robbed people of their basic neural rights. This problem was brought to its extreme in Ken Kesey's novel *One Flew Over the Cuckoo's Nest*, in which the rebellious institutionalized patient Randle McMurphy is punished for bucking authority: he becomes the unlucky recipient of a lobotomy. McMurphy's gleeful personality had unlocked the lives of the other patients in the ward, but the lobotomy turns him into a vegetable. Upon seeing McMurphy's new condition, his docile friend "Chief" Bromden does the favor of suffocating him with a pillow before the other inmates can see the ignominious fate of their leader. Frontal lobotomies, for which Moniz won the Nobel Prize, are no longer considered the proper approach to criminal behavior.[28]

But if the lobotomy stops the crimes, why not do it? The ethical problem pivots on how much a state should be able to change its citizens.* To my mind, this is one of the landmark problems in modern neuroscience: as we come to understand the brain, how can we keep governments from meddling with it? Note that this problem raises its head not just in sensational forms, such as the lobotomy, but in more subtle forms, such as whether second-time sex offenders should be forced to have chemical castration, as they currently are in California and Florida.

But here we propose a new solution, one that can rehabilitate without ethical worries. We call it the prefrontal workout.

THE PREFRONTAL WORKOUT

To help a citizen reintegrate into society, the ethical goal is to change him *as little as possible* to allow his behavior to come into line with society's needs. Our proposal springboards off the knowledge that the brain is a team of rivals, a competition among different

*Incidentally, the lobotomy lost favor not so much because of ethical concerns, but because psychoactive drugs came on the market at the beginning of the 1950s, providing a more expedient approach to the problem.

neural populations. Because it's a competition, this means the outcome can be tipped.

Poor impulse control is a hallmark characteristic of the majority of criminals in the prison system.[29] They generally know the difference between right and wrong actions, and they understand the seriousness of the punishment—but they are hamstrung by an inability to control their impulses. They see a woman with an expensive purse walking alone in an alley, and they cannot think but to take advantage of the opportunity. The temptation overrides the concern for their future.

If it seems difficult to empathize with people who have poor impulse control, just think of all the things you succumb to that you don't want to. Snacks? Alcohol? Chocolate cake? Television? One doesn't have to look far to find poor impulse control pervading our own landscape of decision making. It's not that we don't know what's best for us, it's simply that the frontal lobe circuits representing the long-term considerations can't win the elections when the temptation is present. It's like trying to elect a party of moderates in the middle of war and economic meltdown.

So our new rehabilitative strategy is to give the frontal lobes practice in squelching the short-term circuits. My colleagues Stephen LaConte and Pearl Chiu have begun leveraging real-time feedback in brain imaging to allow this to happen.[30] Imagine that you'd like to get better at resisting chocolate cake. In this experiment, you look at pictures of chocolate cake during brain scanning—and the experimenters determine the regions of your brain involved in the craving. Then the activity in those networks is represented by a vertical bar on a computer screen. Your job is to make the bar go down. The bar acts as a thermometer for your craving: If your craving networks are revving high, the bar is high; if you're suppressing your craving, the bar is low. You stare at the bar and try to make it go down. Perhaps you have insight into what you're doing to resist the cake; perhaps it is inaccessible. In any case, you try out different mental avenues until the bar begins to slowly sink. When it goes down, it means you've successfully recruited frontal circuitry to squelch the activity in the networks involved in impulsive craving. The long term has won over the short. Still looking at pictures of chocolate cake, you practice making the bar go down over and over until you've strengthened those frontal circuits. By this method, you're able to visualize the activity in the parts of your brain that need modulation, and you can witness the effects of different mental approaches you might take.

Returning to the democratic team-of-rivals analogy, the idea is to get a good system of checks and balances into place. This prefrontal workout is designed to level the playing field for debate among the parties, cultivating reflection before action.

And really, that's all maturation is. The main difference between teenage and adult brains is the development of the frontal lobes. The human prefrontal cortex does not fully develop until the early twenties, and this underlies the impulsive behavior of teenagers. The frontal lobes are sometimes called the organ of socialization, because becoming socialized is nothing but developing circuitry to squelch our basest impulses.

This explains why damage to the frontal lobes unmasks unsocialized behavior that we would never have thought was fenced in there. Recall the patients with frontotemporal dementia who shoplift, expose themselves, urinate in public, and burst out into song at inappropriate times. Those zombie systems have been lurking under the surface the whole time, but they've been masked by a normally functioning frontal lobe. The same sort of unmasking happens when a person goes out and gets rip-roaring drunk on a Saturday night: they're disinhibiting normal frontal function and letting the zombies climb onto the main stage.

After training at the prefrontal gym, you might still crave the chocolate cake, but you'll know how to win over the craving instead of letting it win over you. It's not that we don't want to enjoy our impulsive thoughts (*Mmm, cake*), it's merely that we want to endow the frontal cortex with some control over whether we act upon them (*I'll pass*). Similarly, if a person considers committing a criminal act, that's permissible as long as he doesn't take action. For the pedophile, we cannot hope to control whether he is attracted to children. As long as he never acts on it, that may be the best we can hope for as a society that respects individual rights and freedom of thought. We cannot restrict what people think; nor should a legal system hope to set that as its goal. Social policy can only hope to prevent impulsive thoughts from tipping into behavior until they are reflected upon by a healthy neurodemocracy.

Although real-time feedback involves cutting-edge technology, that should not distract from the simplicity of the goal: to enhance a person's capacity for long-term decision making. The goal is to give more control to the neural populations that care about long-term consequences. To inhibit impulsivity. To encourage reflection. If a citizen thinks about long-term consequences and still decides to move forward with an illegal act, then we'll deal with those consequences accordingly. This approach has ethical importance and libertarian appeal. Unlike a lobotomy, which sometimes leaves the patient with only an infantile mentality, this approach opens an opportunity for a willing person to help himself. Instead of a government mandating a psychosurgery, here a government can offer a helping hand to better self-reflection and socialization. This approach leaves the brain intact—no drugs or surgery—and leverages the natural mechanisms of brain plasticity to help the brain help itself. It's a tune-up rather than a product recall.

Not all people who increase their capacity for self-reflection will come to the same sound conclusions, but at least the opportunity to listen to the debate of the neural parties is available. Note also that this approach might restore a bit of the hoped-for power of deterrence, which can work only for people who think about and act upon long-term consequences. For the impulsive, threats of punishment have no real chance to weigh in.

The science of the prefrontal workout is at its very earliest stages, but we have hope that the approach represents the correct model:

it is simultaneously well grounded in biology and ethics, and it allows a person to help himself to better long-term decision making. Like any scientific attempt, it could fail for any number of unforeseen reasons. But at least we have reached a point where we can develop new ideas rather than assuming that incarceration is the only practical solution.

One of the challenges to implementing new rehabilitative approaches is winning popular acceptance. Many people (but not all) have a strong retributive impulse: they want to see punishment, not rehabilitation.[31] I understand that impulse, because I have it too. Every time I hear about a criminal committing an odious act, it makes me so angry that I want to take vigilante-style revenge. But just because we have the drive for something doesn't make it the best approach.

Take xenophobia, the fear of foreigners. It's completely natural. People prefer people who look and sound like them; although contemptible, it is common to dislike outsiders. Our social policies work to cement into place the most enlightened ideas of humanity to surmount the basest facets of human nature. And so the United States passed antidiscrimination housing laws in the form of Title VIII of the Civil Rights Act of 1968. It took a long time to get there, but the fact that we did demonstrates that we are a flexible society that can improve our standards based on better understanding.

And so it goes with vigilantism: despite our understanding of the retributive impulse, we agree to resist it as a society because we know that people can get confused about the facts of a crime, and that everyone deserves the presumption of innocence until proven guilty before a jury of peers. Similarly, as we come to understand more about the biological basis of behavior, it will make sense to subjugate our intuitive notions of blameworthiness in deference to a more constructive approach. We're capable of learning better ideas, and the job of the legal system is to take the very best ideas and carefully mortar them into place to withstand the forces of changing opinion. While brain-based social policy seems distant today, it may not be for long. And it may not always seem counterintuitive.

THE MYTH OF HUMAN EQUALITY

There are more reasons to understand how brains lead to behavior. Along any axis that we measure human beings, we discover a wide-ranging distribution, whether in empathy, intelligence, swimming ability, aggressiveness, or inborn talent at cello or chess.[32] People are not created equal. Although this variability is often imagined to be an issue best swept under the rug, it is in fact the engine of evolution. In each generation, nature tries out as many varieties as it can generate, along all available dimensions—and the products best suited for the environment get to reproduce themselves. For the past billion years this has been a tremendously successful approach, yielding human beings in rocket ships from single self-replicating molecules in pre-biotic soup.

But this variation is also a source of trouble for the legal system, which is built partially upon the premise that humans are all equal before the law. This built-in myth of human equality suggests that all people are equally capable of decision making, impulse control, and comprehending consequences. While admirable, the notion is simply not true.

Some argue that even though the myth may be bullet-riddled, it may still be useful to hold on to. The argument suggests that whether or not the equality is realistic, it yields a "particularly admirable kind of social order, a counterfactual that pays dividends in fairness and stability."[33] In other words, assumptions can be provably wrong and still have utility.

I disagree. As we have seen throughout the book, people do not arrive at the scene with the same capacities. Their genetics and their personal histories mold their brains to quite different end points. In fact, the law partially acknowledges this, because the strain is too great to pretend that *all* brains are equal. Consider age. Adolescents command different skills in decision making and impulse control than do adults; a child's brain is simply not like an adult's brain.[34] So American law draws a bright line between seventeen years and eighteen years to ham-handedly acknowledge this. And the United States Supreme Court ruled in *Roper v Simmons* that those under the age of eighteen when they committed a crime could not be given the death penalty.[35] The law also recognizes that IQ matters. Thus, the Supreme Court made a similar decision that the mentally retarded cannot be executed for capital crimes.

So the law already recognizes that all brains are not created equal. The problem is that the current version of the law uses crude divisions: If you're eighteen we can kill you; if you're one day shy of your eighteenth birthday you're safe. If your IQ is 70, you get the electric chair; if it's 69, get comfortable on your prison mattress. (Because IQ scores fluctuate on different days and with different testing conditions, you'd better hope for the right circumstances if you're near the borderline.)

There's no point in pretending that all non-minor, non–mentally-retarded citizens are equal to one another, because they're not. With different genes and experience, people can be as different on the inside as they are on the outside. As neuroscience improves, we will have a better ability to understand people along a spectrum, rather than in crude, binary categories. And this will allow us to tailor sentencing and rehabilitation for the individual rather than maintain the pretense that all brains respond to the same incentives and deserve the same punishments.

SENTENCING BASED ON MODIFIABILITY

Personalization of the law can go in many directions; I'll suggest one here. Let's return to the case of your daughter writing with a crayon on the wall. In one scenario, she's doing it mischievously; in the other, while she's sleepwalking. Your intuition tells you that you would punish only for the awake case and not for the asleep case. But why? I propose that your intuition may incorporate an important insight about the purpose of punishment. In this case, what matters is not so much your intuition about blameworthiness (although she is clearly not blameworthy when she's asleep), but instead about modifiability. The idea would be to punish only when the behavior is *modifiable*. She cannot modify her behavior in the case of sleepwalking, and therefore punishment would be cruel and fruitless.

I speculate that someday we will be able to base punishment decisions on neuroplasticity. Some people have brains that are better able to respond to classical conditioning (punishment and reward), while other people—because of psychosis, sociopathy, frontal maldevelopment, or other problems—are refractory to change. Take a punishment such as a harsh sentence of breaking rocks: if this is meant to disincentivize prisoners from returning, there is no purpose of this punishment where there is not appropriate brain plasticity to receive it. If there is hope of using classical conditioning to effect a change in behavior that would allow social reintegration, then punishment is appropriate. When a convicted criminal is not going to be usefully changed by punishment, he should simply be warehoused.

Some philosophers have suggested that punishment could be based on the number of options that were available to an actor. A fly, say, is neurally incapable of navigating complex choices, whereas a human (and especially a smart human) has many choices and therefore more control. A system of punishment could be devised, then, in which the degree of punishment goes with the degree of options available to the agent. But I don't think this is the best approach, because someone might have few options but be nonetheless modifiable. Take the non-housetrained puppy. It does not even consider whining and pawing at the door when it has to urinate; the choice was not its to make, because it had not developed the notion of that option. Nonetheless, you scold the dog to modify its central nervous system for appropriate behavior. The same goes for a child who shoplifts.

She does not understand the issues of ownership and economics at first. You punish her not because you feel she had plenty of options, but instead because you understand her to be modifiable. You are doing her a favor: you are socializing her.

This proposal seeks to align punishment with neuroscience. The idea is to replace folk intuitions about blameworthiness with a fairer approach. Although it would be expensive now, societies in the future might experimentally derive an index to measure neuroplasticity— that is, the capacity to modify the circuitry. For those who are modifiable, such as a teenager who still needs further frontal development, a harsh punishment (breaking rocks all summer) would be appropriate. But someone with frontal lobe damage, who will never develop the capacity for socialization, should be incapacitated by the state in a different sort of institution. The same goes for the mentally retarded or schizophrenic; punitive action may slake bloodlust for some, but there is no point in it for society more broadly.

*   *   *

We've spent the first five chapters exploring the degree to which we are not the ones driving the boat. We saw that people have little capacity to choose or explain their actions, motivations, and beliefs, and that the captain's wheel is steered by the unconscious brain, shaped by innumerable generations of evolutionary selection and a lifetime of experiences. The present chapter has explored the social consequences of that: How does the inaccessibility of the brain matter at the level of society? How does it navigate the way we think about blameworthiness, and what should we do about people who behave very differently?

Currently, when a criminal stands in front of the judge's bench, the legal system asks, *Is this person blameworthy?* In the case of Whitman or Alex or a Tourette's patient or a sleepwalker, the system says no. But if you have no obvious biological problem, the system says yes. This cannot be a sensible way to structure a legal system, given the certainty that technology will continue to improve every year and move the position of the "fault" line. It is perhaps too early to say whether every aspect of human behavior will someday be understood as beyond our volition. But in the meantime, the march of science will continue to push the place where we draw our line on the spectrum between volition and non-volition.

As director of Baylor College of Medicine's Initiative on Neuroscience and Law, I have gone around the world lecturing on these issues. The biggest battle I have to fight is the misperception that an improved biological understanding of people's behaviors and internal differences means we will forgive criminals and no longer take them off the streets. That's incorrect. Biological explanation will not exculpate criminals. Brain science will improve the legal system, not impede its function.[36] For the smooth operation of society, we will still remove from the streets those criminals who prove themselves to be over-aggressive, under-empathetic, and poor at controlling their impulses. They will still be taken into the care of the government.

But the important change will be in the *way* we punish the vast range of criminal acts—in terms of rational sentencing and new ideas for rehabilitation. The emphasis will shift from punishment to recognizing problems (both neural and social) and meaningfully addressing them.[37] As one example, we learned in this chapter how the team-of-rivals framework can offer new hope in terms of a rehabilitative strategy.

Further, as we come to better understand the brain, we can concentrate on building societal incentives to encourage good behavior and discourage bad behavior. Effective law requires effective behavioral models: understanding not just how we would *like* people to behave, but how they *actually* behave. As we mine the relationships among neuroscience, economics, and decision making, social policy can be better structured to more effectively leverage these findings.[38] This will reduce our emphasis on retribution in exchange for proactive, preventative policy making.

My argument in this chapter has not been to redefine blameworthiness; instead it is to remove it from the legal argot.

Blameworthiness is a backward-looking concept that demands the impossible task of untangling the hopelessly complex web of genetics and environment that constructs the trajectory of a human life. Consider, for example, that all known serial murderers were abused as children.[39] Does this make them less blameworthy? Who cares? It's the wrong question to ask. The knowledge that they were abused encourages us to prevent child abuse, but it does nothing to change the way we deal with the particular serial murderer standing in front of the bench. We still need to warehouse him. We need to keep him off the streets, irrespective of his past misfortunes. The child abuse cannot serve as a meaningful biological excuse; the judge must take action to keep society safe.

The concept and word to replace *blameworthiness* is *modifiability*, a forward-looking term that asks, What can we do from here? Is rehabilitation available? If so, great. If not, will the punishment of a prison sentence modify future behavior? If so, send him to prison. If punishment won't help, then take the person under state control for the purposes of incapacitation, not retribution.

My dream is to build an evidence-based, neurally compatible social policy instead of one based on shifting and provably bad intuitions. Some people wonder whether it's unfair to take a scientific approach to sentencing—after all, where's the humanity there? But this concern should always be met with a question: what's the alternative? As it stands now, ugly people receive longer sentences than attractive people; psychiatrists have no capacity to guess which sex offenders will reoffend; and our prisons are overcrowded with drug addicts who could be more usefully dealt with by rehabilitation rather than incarceration. So is current sentencing really better than a scientific, evidence-based approach?

Neuroscience is just beginning to scratch the surface of questions that were once only in the domain of philosophers and psychologists, questions about how people make decisions and whether they are truly "free." These are not idle questions, but ones that will shape the future of legal theory and the dream of a biologically informed jurisprudence.[40]

7

Life After the Monarchy

"As for men, those myriad little detached ponds with their own
swarming corpuscular life, what were they but a way that water has
of going about beyond the reach of rivers?"
—Loren Eiseley, "The Flow of the River", *The Immense Journey*

FROM DETHRONEMENT TO DEMOCRACY

After Galileo discovered the moons of Jupiter in his homemade telescope in 1610, religious critics decried his new sun-centered theory
as a dethronement of man. They didn't suspect that this was only
the first dethronement of several. One hundred years later, the study
of sedimentary layers by the Scottish farmer James Hutton toppled
the Church's estimate of the age of the Earth—making it eight
hundred thousand times older. Not long afterward, Charles Darwin
relegated humans to just another branch in the swarming animal
kingdom. At the beginning of the 1900s, quantum mechanics
irreparably altered our notion of the fabric of reality. In 1953,
Francis Crick and James Watson deciphered the structure of DNA,
replacing the mysterious ghost of life with something that we can
write down in sequences of four letters and store in a computer.

And over the past century, neuroscience has shown that the
conscious mind is not the one driving the boat. A mere four hundred
years after our fall from the center of universe, we have experienced the fall from the center of ourselves. In the first chapter we
saw that conscious access to the machinery under the hood is slow,
and often doesn't happen at all. We then learned that the way we see
the world is not necessarily what's out there: vision is a construction

of the brain, and its only job is to generate a useful narrative at our scales of interactions (say, with ripe fruits, bears, and mates). Visual illusions reveal a deeper concept: that our *thoughts* are generated by machinery to which we have no direct access. We saw that useful routines become burned down into the circuitry of the brain, and that once they are there, we no longer have access to them. Instead, consciousness seems to be about setting goals for what should be burned into the circuitry, and it does little beyond that. In Chapter 5 we learned that minds contain multitudes, which explains why you can curse at yourself, laugh at yourself, and make contracts with yourself. And in Chapter 6 we saw that brains can operate quite differently when they are changed by strokes, tumors, narcotics, or any variety of events that alter the biology. This agitates our simple notions of blameworthiness.

In the wake of all the scientific progress, a troubling question has surfaced in the minds of many: what is left for humans after all these dethronements? For some thinkers, as the immensity of the universe became more apparent, so did humankind's inconsequentiality—we began to dwindle in importance virtually to the vanishing point. It became clear that the epochal time scales of civilizations represented only a flash in the long history of multicellular life on the planet, and the history of life is only a flash in the history of the planet itself. And that planet, in the vastness of the universe, is only a tiny speck of matter floating away from other specks at cosmic speed through the desolate curvature of space. Two hundred million years from now, this vigorous, productive planet will be consumed in the expansion of the sun. As Leslie Paul wrote in *Annihilation of Man*:

> All life will die, all mind will cease, and it will all be as if it had never happened. That, to be honest, is the goal to which evolution is traveling, that is the "benevolent" end of the furious living and furious dying. . . . All life is no more than a match struck in the dark and blown out again. The final result . . . is to deprive it completely of meaning.[1]

After building many thrones and falling from all of them, man looked around; he wondered whether he had accidentally been generated in a blind and purposeless cosmic process, and he strove to salvage some sort of purpose. As the theologian E. L. Mascall wrote:

> The difficulty which civilized Western man in the world today experiences is in convincing himself that he has any special assigned status in the universe. . . . Many of the psychological disorders which are so common and distressing a feature of our time are, I believe, to be traced to this cause.[2]

Philosophers such as Heidegger, Jaspers, Shestov, Kierkegaard, and Husserl all scrambled to address the meaninglessness with which the dethronements seemed to have left us. In his 1942 book *Le mythe de Sisyphe*, Albert Camus introduced his philosophy of the absurd, in which man searches for meaning in a fundamentally meaningless world. In this context, Camus proposed that the only real question in philosophy is whether or not to commit suicide. (He concluded that one should *not* commit suicide; instead, one should live to revolt against the absurd life, even though it will always be without hope. It is possible that he was forced to this conclusion because the opposite would have impeded sales of his book unless he followed his own prescription—a tricky catch-22.)

I suggest that the philosophers may have been taking the news of the dethronements a bit too hard. Is there really nothing left for mankind after all these dethronements? The situation is likely to be the opposite: as we plumb further down, we will discover ideas much broader than the ones we currently have on our radar screens, in the same way that we have begun to discover the gorgeousness of the microscopic world and the incomprehensible scale of the cosmos. The act of dethronement tends to open up something bigger than us, ideas more wonderful than we had originally imagined. Each discovery taught us that reality far outstrips human imagination and guesswork. These advances deflated the power of intuition and tradition as an oracle of our future, replacing them with more productive ideas, bigger realities, and new levels of awe.

In the case of Galileo's discovery that we are not at the center of the universe, we now know something much greater: that our solar system is one of billions of trillions. As I mentioned earlier, even if life emerges only on one planet in a billion, it means there may be millions and millions of planets teeming with activity in the cosmos. To my mind, that's a bigger and brighter idea than sitting at a lonely center surrounded by cold and distant astral lamps. The dethronement led to a richer, deeper understanding, and what we lost in egocentrism was counterbalanced in surprise and wonder.

Similarly, understanding the age of the Earth opened previously unimaginable time vistas, which in turn opened the possibility of understanding natural selection. Natural selection is used daily in laboratories around the globe to select colonies of bacteria in research to combat disease. Quantum mechanics has given us the transistor (the heart of our electronics industry), lasers, magnetic resonance imaging, diodes, and memory in USB flash drives—and may soon deliver the revolutions of quantum computing, tunneling, and teleportation. Our understanding of DNA and the molecular basis of inheritance has allowed us to target disease in ways that were unimaginable a half century ago. By taking seriously the discoveries of science, we have eradicated smallpox, traveled to the moon, and launched the information revolution. We have tripled life spans, and by targeting diseases at the molecular level, we will soon float the average life span beyond one hundred years. Dethronements often equal progress.

In the case of the dethronement of the conscious mind, we gain better inroads to understand human behavior. Why do we find things beautiful? Why are we bad at logic? Who's cursing at whom when we get mad at ourselves? Why do people fall for the allure of adjustable-rate mortgages? How can we steer a car so well but find ourselves unable to describe the process?

This improved understanding of human behavior can translate directly into improved social policy. As one example, an understanding of the brain matters for structuring incentives. Recall the fact from Chapter 5 that people negotiate with themselves, making an endless series of Ulysses contracts. This leads to ideas like the proposed diet plan from that chapter: people who want to lose weight can deposit a good deal of money into an escrow holding. If they meet their weight-loss goal by a specified deadline, they get the money back; otherwise they lose it all. This structure allows people in a moment of sober reflection to recruit support against their short-term decision making—after all, they know that their future self will be tempted to eat with impunity. Understanding this aspect of human nature allows this sort of contract to be usefully introduced in various settings—for example, getting an employee to siphon a little portion of his monthly paycheck into an individual retirement account. By making the decision up front, he can avoid the temptation of spending later.

Our deeper understanding of the inner cosmos also gives us a clearer view of philosophical concepts. Take virtue. For millennia, philosophers have been asking what it is and what we can do to enhance it. The team-of-rivals framework gives new inroads here. We can often interpret the rivalrous elements in the brain as analogous to *engine* and *brakes*: some elements are driving you toward a behavior while others are trying to stop you. At first blush, one might think virtue consists of not wanting to do bad things. But in a more nuanced framework, a virtuous person can have strong lascivious drives as long as he also commands sufficient braking power to surmount them. (It is also the case that a virtuous actor can have minimal temptations and therefore no requirement for good brakes, but one could suggest that the more virtuous person is he who has fought a stronger battle to resist temptation rather than he who was never enticed.) This sort of approach is possible only when we have a clear view of the rivalry under the hood, and not if we believe people possess only a single mind (as in *mens rea,* "the guilty mind"). With the new tools, we can consider a more nuanced battle between different brain regions and how the battle tips. And that opens up new opportunities for rehabilitation in our legal system: when we understand how the brain is really operating and why impulse control fails in some fraction of the population, we can develop direct new strategies to strengthen long-term decision making and tip the battle in its favor.

Additionally, an understanding of the brain has the potential to elevate us to a more enlightened system of sentencing. As we saw in the previous chapter, we will be able to replace the problematic concept of blameworthiness with a practical, future-looking corrections system (*What is this person likely to do from here?*) instead of a retrospective one (*How much was it his fault?*). Someday the legal system may be able to approach neural and behavioral problems in the same manner that medicine studies lung or bone problems. Such biological realism will not clear criminals, but instead will introduce rational sentencing and customized rehabilitation by adopting a prospective approach instead of a retrospective one.

A better understanding of neurobiology may lead to better social policy. But what does it mean for understanding our own lives?

KNOWING THYSELF

> "Know then thyself, presume not God to scan. The proper study of mankind is man."
>
> —Alexander Pope

On February 28, 1571, on the morning of his thirty-eighth birthday, the French essayist Michel de Montaigne decided to make a radical change in his life's trajectory. He quit his career in public life, set up a library with one thousand books in a tower at the back of his large estate, and spent the rest of his life writing essays about the complex, fleeting, protean subject that interested him the most: *himself.* His first conclusion was that a search to know oneself is a fool's errand, because the self continuously changes and keeps ahead of a firm description. That didn't stop him from searching, however, and his question has resonated through the centuries: *Que sais-je?* (What do I know?)

It was, and remains, a good question. Our exploration of the inner cosmos certainly disabuses us of our initial, uncomplicated, intuitive notions of knowing ourselves. We see that self-knowledge requires as much work from the outside (in the form of science) as from the inside (introspection). This is not to say that we cannot grow better at introspection. After all, we can learn to pay attention to what we're really seeing out there, as a painter does, and we can attend more closely to our internal signals, as a yogi does. But there are limits to introspection. Just consider the fact that your peripheral nervous system employs one hundred million neurons to control the activities in your gut (this is called the enteric nervous system). One hundred million neurons, and no amount of your introspection can touch this. Nor, most likely, would you want it to. It's better off running as the automated, optimized machinery that it is, routing food along your gut and providing chemical signals to control the digestion factory without asking your opinion on the matter.

Beyond lack of access, there could even be prevention of access. My colleague Read Montague once speculated that we might have algorithms that protect us from ourselves. For example, computers have boot sectors which are inaccessible by the operating system— they are too important for the operation of the computer for any other higher level systems to find inroads and gain admission, under any circumstances. Montague noted that whenever we try to think about ourselves too much, we tend to "blink out"—and perhaps this is because we are getting too close to the boot sector. As Ralph Waldo Emerson wrote over a century earlier, "Everything intercepts us from ourselves."

Much of who we are remains outside our opinion or choice. Imagine trying to change your sense of beauty or attraction. What would happen if society asked you to develop and maintain an attraction to someone of the gender to which you are currently not attracted? Or someone well outside the age range to which you are currently attracted? Or outside your species? Could you do it? Doubtful. Your most fundamental drives are stitched into the fabric of your neural circuitry, and they are inaccessible to you. You find certain things more attractive than others, and you don't know why.

Like your enteric nervous system and your sense of attraction, almost the entirety of your inner universe is foreign to you. The ideas that strike you, your thoughts during a daydream, the bizarre content of your nightdreams—all these are served up to you from unseen intracranial caverns.

So what does all of this mean for the Greek admonition γνῶθι σεαυτόν—know thyself—inscribed prominently in the forecourt of the Temple of Apollo at Delphi? Can we ever know ourselves more deeply by studying our neurobiology? Yes, but with some caveats. In the face of the deep mysteries presented by quantum physics, the physicist Niels Bohr once suggested that an understanding of the structure of the atom could be accomplished only by changing the definition "to understand." One could no longer draw pictures of an atom, true, but instead one could now predict experiments about its behavior out to fourteen decimal places. Lost assumptions were replaced by something richer.

By the same token, to know oneself may require a change of definition of "to know." Knowing yourself now requires the understanding that the conscious *you* occupies only a small room in the mansion of the brain, and that it has little control over the reality constructed for you. The invocation to know thyself needs to be considered in new ways.

Let's say you wanted to know more about the Greek idea of knowing thyself, and you asked me to explain it further. It probably wouldn't be helpful if I said, "Everything you need to know is in the individual letters: γ ν ῶ θ ι σ ε α υ τ ό ν." If you don't read Greek, the elements are nothing but arbitrary shapes. And even if you *do* read Greek, there's so much more to the idea than the letters—instead you would want to know the culture from which it sprung, the emphasis on introspection, the suggestion of a path to enlightenment.[3] Understanding the phrase requires more than learning the letters. And this is the situation we're in when we look at trillions of neurons and their sextillions of voyaging proteins and biochemicals. What does it mean to know ourselves from that totally unfamiliar perspective? As we will see in a moment, we need the neurobiological data, but we also need quite a bit more to know ourselves.

Biology is a terrific approach, but it's limited. Consider lowering a medical scope down your lover's throat while he or she reads poetry to you. Get a good, close-up view of your lover's vocal chords, slimy and shiny, contracting in and out in spasms. You could study this until you were nauseated (maybe sooner rather than later, depending on your tolerance for biology), but it would get you no closer to understanding why you love nighttime pillow talk. By itself, in its raw form, the biology gives only partial insight. It's the best we can do right now, but it's far from complete. Let's turn to this in more detail now.

WHAT IT DOES AND DOESN'T MEAN TO BE CONSTRUCTED OF PHYSICAL PARTS

One of the most famous examples of brain damage comes from a twenty-five-year-old work-gang foreman named Phineas Gage. The *Boston Post* reported on him in a short article on September 21, 1848, under the headline "Horrible Accident":

> As Phineas P. Gage, a foreman on the railroad in Cavendish, was yesterday engaged in tamping for a blast, the powder exploded, carrying an instrument through his head an inch and a fourth in [diameter], and three feet and [seven] inches in length, which he was using at the time. The iron entered on the side of his face, shattering the upper jaw, and passing back of the left eye, and out at the top of the head.

The iron tamping rod clattered to the ground twenty-five yards away. While Gage wasn't the first to have his skull punctured and a portion of his brain spirited away by a projectile, he was the first to not die from it. In fact, Gage did not even lose consciousness.

The first physician to arrive, Dr. Edward H. Williams, did not believe Gage's statement of what had just happened, but instead "thought he [Gage] was deceived." But Williams soon understood the gravity of what had happened when "Mr. G. got up and vomited; the effort of vomiting pressed out about half a teacupful of the brain, which fell upon the floor."

The Harvard surgeon who studied his case, Dr. Henry Jacob Bigelow, noted that "the leading feature of this case is its improbability. . . . [It is] unparalleled in the annals of surgery."[4] The *Boston Post* article summarized this improbability with just one more sentence: "The most singular circumstance connected with this melancholy affair is that he was alive at 2:00 this afternoon, and in full possession of his reason, and free from pain."[5]

Gage's survival alone would have made an interesting medical case; it became a famous case because of something else that came to light. Two months after the accident his physician reported that Gage was "feeling better in every respect . . . walking about the house again; says he feels no pain in the head." But foreshadowing a larger problem, the doctor also noted that Gage "appears to be in a way of recovering, if he can be controlled."

What did he mean, "if he can be controlled"? It turned out that the preaccident Gage had been described as "a great favorite" among his team, and his employers had hailed him as "the most efficient and capable foreman in their employ." But after the brain change, his employers "considered the change in his mind so marked that they could not give him his place again." As Dr. John Martyn Harlow, the physician in charge of Gage, wrote in 1868:

> The equilibrium or balance, so to speak, between his intellectual faculties and animal propensities, seems to have been destroyed. He is fitful, irreverent, indulging at times in the grossest profanity (which was not previously his custom), manifesting but little deference for his fellows, impatient of restraint or advice when

it conflicts with his desires, at times pertinaciously obstinate, yet capricious and vacillating, devising many plans of future operations, which are no sooner arranged than they are abandoned in turn for others appearing more feasible. A child in his intellectual capacity and manifestations, he has the animal passions of a strong man. Previous to his injury, although untrained in the schools, he possessed a well-balanced mind, and was looked upon by those who knew him as a shrewd, smart businessman, very energetic and persistent in executing all his plans of operation. In this regard his mind was radically changed, so decidedly that his friends and acquaintances said he was "no longer Gage."[6]

In the intervening 143 years we have witnessed many more of nature's tragic experiments—strokes, tumors, degeneration, and every variety of brain injury—and these have produced many more cases like Phineas Gage's. The lesson from all these cases is the same: the condition of your brain is central to who you are. The *you* that all your friends know and love cannot exist unless the transistors and screws of your brain are in place. If you don't believe this, step into any neurology ward in any hospital. Damage to even small parts of the brain can lead to the loss of shockingly specific abilities: the ability to name animals, or to hear music, or to manage risky behavior, or to distinguish colors, or to arbitrate simple decisions. We've already seen examples of this with the patient who lost the ability to see motion (Chapter 2), and the Parkinson's gamblers and frontotemporal shoplifters who lost the ability to manage risk-taking (Chapter 6). Their essence was changed by the changes in their brain.

All of this leads to a key question: do we possess a soul that is separate from our physical biology—or are we simply an enormously complex biological network that mechanically produces our hopes, aspirations, dreams, desires, humor, and passions?[7] The majority of people on the planet vote for the extrabiological soul, while the majority of neuroscientists vote for the latter: an essence that is a natural property that emerges from a vast physical system, and nothing more besides. Do we know which answer is correct? Not with certainty, but cases like Gage's certainly seem to weigh in on the problem.

The *materialist* viewpoint states that we are, fundamentally, made only of physical materials. In this view, the brain is a system whose operation is governed by the laws of chemistry and physics— with the end result that all of your thoughts, emotions, and decisions are produced by natural reactions following local laws to lowest potential energy. We are our brain and its chemicals, and any dialing of the knobs of your neural system changes *who you are*. A common version of materialism is called *reductionism*; this theory puts forth the hope that we can understand complex phenomena like happiness, avarice, narcissism, compassion, malice, caution, and awe by successively *reducing* the problems down to their small-scale biological pieces and parts.

At first blush, the reductionist viewpoint sounds absurd to many people. I know this because I ask strangers their opinion about it when I sit next to them on airplanes. And they usually say something like "Look, all that stuff—how I came to love my wife, why I chose my job, and all the rest—that has nothing to do with the chemistry of my *brain*. It's just *who I am*." And they're right to think that the connection between your essence as a person and a squishy confederacy of cells seems distant at best. The passengers' decisions came from *them*, not a bunch of chemicals cascading through invisibly small cycles. Right?

But what happens when we crash into enough cases like Phineas Gage's? Or when we turn the spotlight on other influences on the brain—far more subtle than a tamping rod—that change people's personalities?

Consider the powerful effects of the small molecules we call narcotics. These molecules alter consciousness, affect cognition, and navigate behavior. We are slave to these molecules. Tobacco, alcohol, and cocaine are self-administered universally for the purpose of mood changing. If we knew nothing else about neurobiology, the mere existence of narcotics would give us all the evidence we require that our behavior and psychology can be commandeered at the molecular level. Take cocaine as an example. This drug interacts with a specific network in the brain, one that registers rewarding events—anything from slaking your thirst with a cool iced tea, to winning a smile from the right person, to cracking a tough problem, to hearing "Good job!" By tying positive outcomes to the behaviors that led to them, this widespread neural circuit (known as the mesolimbic dopamine system) learns how to optimize behavior in the world. It aids us in getting food, drink, and mates, and it helps us navigate life's daily decisions.*

Out of context, cocaine is a totally uninteresting molecule: seventeen carbon atoms, twenty-one hydrogens, one nitrogen, and four oxygens. What makes cocaine *cocaine* is the fact that its accidental shape happens to fit lock-and-key into the microscopic machinery of the reward circuits. The same goes for all four major classes of drugs of abuse: alcohol, nicotine, psychostimulants (such as amphetamines), and opiates (such as morphine): by one inroad or another, they all plug into this reward circuitry.[8] Substances that can give a shot in the arm to the mesolimbic dopamine system have self-reinforcing effects, and users will rob stores and mug elderly people to continue obtaining these specific molecular shapes. These chemicals, working their magic at scales one thousand times smaller than the width of a human hair, make the users feel invincible and euphoric. By plugging into the dopamine system, cocaine and its cousins commandeer the reward system, telling the brain that this is the best possible thing that could be happening. The ancient circuits are hijacked.

The cocaine molecules are hundreds of millions of times smaller than the tamping rod that shot through Phineas Gage's brain, and yet the lesson is the same: who you are depends on the sum total of your neurobiology.

---

*The basic architecture of this reward circuit is highly conserved throughout evolution. The brain of a honeybee uses the same reward programs that your brain does, running the same software program on a much more compact piece of hardware. (See Montague, et al., "Bee foraging.").

And the dopamine system is only one of hundreds of examples. The exact levels of dozens of other neurotransmitters—for example, serotonin—are critical for who you believe yourself to be. If you suffer from clinical depression, you will probably be prescribed a medication known as a selective serotonin reuptake inhibitor (abbreviated as an SSRI)—something such as fluoxetine or sertraline or paroxetine or citalopram. Everything you need to know about how these drugs work is contained in the words "uptake inhibitor": normally, channels called transporters take up serotonin from the space between neurons; the inhibition of these channels leads to a higher concentration of serotonin in the brain. And the increased concentration has direct consequences on cognition and emotion. People on these medications can go from crying on the edge of their bed to standing up, showering, getting their job back, and rescuing healthy relationships with the people in their life. All because of a subtle fine-tuning of a neurotransmitter system.[9] If this story weren't so common, its bizarreness could be more easily appreciated.

It's not just neurotransmitters that influence your cognition. The same goes for hormones, the invisibly small molecules that surf the bloodstream and cause commotion at every port they visit. If you inject a female rat with estrogen, she will begin sexual seeking; testosterone in a male rat causes aggression. In the previous chapter we learned about the wrestler Chris Benoit, who took massive doses of testosterone and murdered his wife and his own child in a hormone rage. And in Chapter 4 we saw that the hormone vasopressin is linked to fidelity. As another example, just consider the hormone fluctuations that accompany normal menstrual cycles. Recently, a female friend of mine was at the bottom of her menstrual mood changes. She put on a wan smile and said, "You know, I'm just not myself for a few days each month." Being a neuroscientist, she then reflected for a moment and added, "Or maybe *this* is the real me, and I'm actually someone else the other twenty-seven days of the month." We laughed. She was not afraid to view herself as the sum total of her chemicals at any moment. She understood that what we think of as *her* is something like a time-averaged version.

All this adds up to something of a strange notion of a self. Because of inaccessible fluctuations in our biological soup, some days we find ourselves more irritable, humorous, well spoken, calm, energized, or clear-thinking. Our internal life and external actions are steered by biological cocktails to which we have neither immediate access nor direct acquaintance.

And don't forget that the long list of influences on your mental life stretches far beyond chemicals—it includes the details of circuitry, as well. Consider epilepsy. If an epileptic seizure is focused in a particular sweet spot in the temporal lobe, a person won't have motor seizures, but instead something more subtle. The effect is something like a cognitive seizure, marked by changes of personality, hyperreligiosity (an obsession with religion and a feeling of religious certainty), hypergraphia (extensive writing on a subject, usually about religion), the false sense of an external presence, and, often, the hearing of voices that are attributed to a god.[10] Some fraction of history's prophets, martyrs, and leaders appear to have had temporal lobe epilepsy.[11] Consider Joan of Arc, the sixteen-year-old-girl who managed to turn the tide of the Hundred Years War because she believed (and convinced the French soldiers) that she was hearing voices from Saint Michael the archangel, Saint Catherine of Alexandria, Saint Margaret, and Saint Gabriel. As she described her experience, "When I was thirteen, I had a voice from God to help me to govern myself. The first time, I was terrified. The voice came to me about noon: it was summer, and I was in my father's garden." Later she reported, "Since God had commanded me to go, I must do it. And since God had commanded it, had I had a hundred fathers and a hundred mothers, and had I been a king's daughter, I would have gone." Although it's impossible to retrospectively diagnose with certainty, her typical reports, increasing religiosity, and ongoing voices are certainly consistent with temporal lobe epilepsy. When brain activity is kindled in the right spot, people hear voices. If a physician prescribes an anti-epileptic medication, the seizures go away and the voices disappear. Our reality depends on what our biology is up to.

Influences on your cognitive life also include tiny nonhuman creatures: microorganisms such as viruses and bacteria hold sway over behavior in extremely specific ways, waging invisible battles inside us. Here's my favorite example of a microscopically small organism taking over the behavior of a giant machine: the rabies virus. After a bite from one mammal to another, this tiny bullet-shaped virus climbs its way up the nerves and into the temporal lobe of the brain. There it ingratiates itself into the local neurons, and by changing the local patterns of activity it induces the infected host to aggression, rage, and a propensity to bite. The virus also moves into the salivary glands, and in this way it is passed on through the bite to the next host. By steering the behavior of the animal, the virus ensures its spread to other hosts. Just think about that: the virus, a measly seventy-five billionths of a meter in diameter, survives by commandeering the massive body of an animal twenty-five million times larger than it. It would be like you finding a creature 28,000 miles tall and doing something very clever to bend its will to yours.[12] The critical take-home lesson is that invisibly small changes inside the brain can cause massive changes to behavior. Our choices are inseparably married to the tiniest details of our machinery.[13]

As a final example of our dependence on our biology, note that tiny mutations in single genes also determine and change behavior. Consider Huntington's disease, in which creeping damage in the frontal cortex leads to changes in personality, such as aggressiveness, hypersexuality, impulsive behavior, and disregard for social norms—all happening years before the more recognizable symptom of spastic limb movement appears.[14] The point to appreciate is that Huntington's is caused by a mutation in a single gene. As Robert Sapolsky summarizes it, "Alter one gene among tens of thousands and, approximately halfway through one's life, there occurs a dramatic transformation of personality."[15] In the face of such examples, can we conclude anything other than a dependence of our essence on the details of our biology? Could you tell a person with Huntington's to use his "free will" to quit acting so strangely?

So we see that the invisibly small molecules we call narcotics, neurotransmitters, hormones, viruses, and genes can place their little hands on the steering wheel of our behavior. As soon as your drink is spiked, your sandwich is sneezed upon, or your genome picks up a mutation, your ship moves in a different direction. Try as you might to make it otherwise, the changes in your machinery lead to changes in you. Given these facts on the ground, it is far from clear that we hold the option of "choosing" who we would like to be. As the neuroethicist Martha Farah puts it, if an anti-depressant pill "can help us take everyday problems in stride, and if a stimulant can help us meet our deadlines and keep our commitments at work, then must not unflabbable temperaments and conscientious characters also be features of people's bodies? And if so, is there anything about people that is *not* a feature of their bodies?"[16]

Who you turn out to be depends on such a vast network of factors that it will presumably remain impossible to make a one-to-one mapping between molecules and behavior (more on that in the moment). Nonetheless, despite the complexity, your world is directly tied to your biology. If there's something like a soul, it is at minimum tangled irreversibly with the microscopic details. Whatever else may be going on with our mysterious existence, our connection to our biology is beyond doubt. From this point of view, you can see why biological reductionism has a strong foothold in modern brain science. But reductionism isn't the whole story.

FROM THE COLOR OF YOUR PASSPORT TO EMERGENT PROPERTIES

Most people have heard of the Human Genome Project, in which our species successfully decoded the billions-of-letters-long sequence in our own genetic codebook. The project was a landmark achievement, hailed with the proper fanfare.

Not everyone has heard that the project has been, in some sense, a failure. Once we sequenced the whole code, we didn't find the hoped-for breakthrough answers about the genes that are unique to humankind; instead we discovered a massive recipe book for building the nuts and bolts of biological organisms. We found that other animals have essentially the same genome we do; this is because they are made of the same nuts and bolts, only in different configurations. The human genome is not terribly different from the frog genome, even though humans are terribly different from frogs. At least, humans and frogs *seem* quite different at first. But keep in mind that both require the recipes to build eyes, spleens, skin, bones, hearts, and so on. As a result, the two genomes are not so dissimilar. Imagine going to different factories and examining the pitches and lengths of the screws used. This would tell you little about the function of the final product—say, a toaster versus a blow dryer. Both have similar elements configured into different functions.

The fact that we didn't learn what we thought we might is not a criticism of the Human Genome Project; it had to be done as a first step. But it *is* to acknowledge that successive levels of reduction are doomed to tell us very little about the questions important to humans.

Let's return to the Huntington's example, in which a single gene determines whether or not you'll develop the disease. That sounds like a success story for reductionism. But note that Huntington's is one of the very few examples that can be dredged up for this sort of effect. The reduction of a disease to a *single* mutation is extraordinarily rare: most diseases are polygenetic, meaning that they result from subtle contributions from tens or even hundreds of different genes. And as science develops better techniques, we are discovering that not just the coding regions of genes matter, but also the areas in between—what used to be thought of as "junk" DNA. Most diseases seem to result from a perfect storm of numerous minor changes that combine in dreadfully complex ways.

But the situation is far worse than just a multiple-genes problem: the contributions from the genome can really be understood only in the context of interaction with the environment. Consider schizophrenia, a disease for which teams of researchers have been gene hunting for decades now. Have they found any genes that correlate with the disease? Sure they have. Hundreds, in fact. Does the possession of any one of these genes offer much in the way of prediction about who will develop schizophrenia as a young adult? Very little. No single gene mutation is as predictive of schizophrenia as the color of your passport.

What does your passport have to do with schizophrenia? It turns out that the social stress of being an immigrant to a new country is one of the critical factors in developing schizophrenia.[17] In studies across countries, immigrant groups who differ the most in culture and appearance from the host population carry the highest risk. In other words, a lower level of social acceptance into the majority correlates with a higher chance of a schizophrenic break. In ways not currently understood, it appears that repeated social rejection perturbs the normal functioning of the dopamine systems. But even these generalizations don't tell the whole story, because within a single immigrant group (say, Koreans in America), those who feel worse about their ethnic differences from the majority are more likely to become psychotic. Those who are proud and comfortable with their heritage are mentally safer.

This news comes as a surprise to many. Is schizophrenia genetic or isn't it? The answer is that genetics play a *role*. If the genetics produce nuts and bolts that have a slightly strange shape, the whole system may run in an unusual manner when put in particular environments. In other environments, the shape of the nuts and bolts may not matter. When all is said and done, how a person turns out depends on much more than the molecular suggestions written down in the DNA.

Remember what we said earlier about having an 828 percent

higher chance of committing a violent crime if you carry the Y chromosome? The statement is factual, but the important question to ask is this: why aren't *all* males criminals? That is, only 1 percent of males are incarcerated.[18] What's going on?

The answer is that knowledge of the genes alone is not sufficient to tell you much about behavior. Consider the work of Stephen Suomi, a researcher who raises monkeys in natural environments in rural Maryland. In this setting, he is able to observe the monkeys' social behavior from their day of birth.[19] One of the first things he noticed was that monkeys begin to express different personalities from a surprisingly early age. He saw that virtually every social behavior was developed, practiced, and perfected during the course of peer play by four to six months of age. This observation would have been interesting by itself, but Suomi was able to combine the behavioral observations with regular blood testing of hormones and metabolites, as well as genetic analysis.

What he found among the baby monkeys was that 20 percent of them displayed social anxiety. They reacted to novel, mildly stressful social situations with unusually fearful and anxious behavior, and this correlated with long-lasting elevations of stress hormones in their blood.

On the other end of the social spectrum, 5 percent of the baby monkeys were overly aggressive. They showed impulsive and inappropriately belligerent behavior. These monkeys had low levels of a blood metabolite related to the breakdown on the neurotransmitter serotonin.

Upon investigation, Suomi and his team found that there were two different "flavors" of genes (called alleles by geneticists) that one could possess for a protein involved in transporting serotonin[20]—let's call these the short and long forms. The monkeys with the short form showed poor control of violence, while those with the long form displayed normal behavioral control.

But that turned out to be only part of the story. How a monkey's personality developed depended on its environment as well. There were two ways the monkeys could be reared: with their mothers (good environment) or with their peers (insecure attachment relationships). The monkeys with the short form ended up as the aggressive type when they were raised with their peers, but did much better when they were raised with their mothers. For those with the long form of the gene, the rearing environment did not seem to matter much; they were well adjusted in either case.

There are at least two ways to interpret these results. The first is that the long allele is a "good gene" that confers resilience against a bad childhood environment (lower left corner of the table below). The second is that a good mothering relationship somehow gives resiliency for those monkeys who would otherwise turn out to be bad seeds (upper right corner). These two interpretations are not exclusive, and they both boil down to the same important lesson: a combination of genetics and environment matters for the final outcome.

	Raised with peers	Raised with mother
Short allele	aggressive	fine
Long allele	fine	fine

With the success of the monkey studies, people began to study gene-environment interactions in humans.[21] In 2001, Avshalom Caspi and his colleagues began to wonder whether there are genes for depression. When they went on the hunt, they found that the answer is "sort of." They learned that there are genes that *predispose* you; whether you actually suffer from depression depends on your life's events.[22] The researchers discovered this by carefully interviewing dozens of people to find out what sort of major traumatic events had transpired in their lives: loss of a loved one, a major car accident, and the like. For each participant, they also analyzed the genetics—specifically, the form of a gene involved in regulation of serotonin levels in the brain. Because people carry two copies of the gene (one from each parent), there are three possible combinations someone might carry: short/short, short/long, or long/long. The amazing result was that the short/short combination predisposed the participants to clinical depression, but only if they experienced an increasing number of bad life events. If they were lucky enough to live a good life, then carrying the short/short combination made them no more likely than anyone else to become clinically depressed. But if they were unlucky enough to run into serious troubles, including events that were entirely out of their control, then they were more than twice as likely to become depressed as someone with the long/long combination.

A second study addressed a deep societal concern: those with abusive parents tend to be abusive themselves. Many people believe this statement, but is it really true? And does it matter what kind of genes the child is carrying? What caught the attention of researchers was the fact that some abused children become violent as adults while others do not. When all the obvious factors were controlled for, the fact stood that childhood abuse,

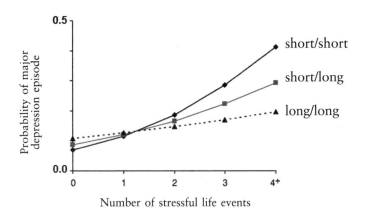

Predispositions in the genes. Why do stressful experiences lead to depression in some individuals but not in others? It may be a matter of genetic predisposition. From Caspi et al., *Science*, 2003.

by itself, did not predict how an individual would turn out. Inspired to understand the difference between those who perpetuate the violence and those who do not, Caspi and his colleagues discovered that a small change in the expression of a particular gene differentiated these children.[23] Children with low expression of the gene were more likely to develop conduct disorders and become violent criminals as adults. However, this bad outcome was much more likely if the children were abused. If they harbored the "bad" forms of the gene but had been spared childhood abuse, they were not likely to become abusers. And if they harbored the "good" forms, then even a childhood of severe maltreatment would not necessarily drive them to continue the cycle of violence.

A third example comes from the observation that smoking cannabis (marijuana) as a teenager increases the probability of developing psychosis as an adult. But this connection is true only for some people, and not for others. By this point, you can guess the punch line: a genetic variation underlies one's susceptibility to this. With one combination of alleles, there is a strong link between cannabis use and adult psychosis; with a different combination, the link is weak.[24]

Similarly, psychologists Angela Scarpa and Adrian Raine measured differences in brain function among people diagnosed with antisocial personality disorder—a syndrome characterized by a total disregard for the feelings and rights of others, and one that is highly prevalent among the criminal population. The researchers found that antisocial personality disorder had the highest likelihood of occurring when brain abnormalities were *combined* with a history of adverse environmental experiences.[25] In other words, if you have certain problems with your brain but are raised in a good home, you might turn out okay. If your brain is fine and your home is terrible, you might still turn out fine. But if you have mild brain damage *and* end up with a bad home life, you're tossing the dice for a very unlucky synergy.

These examples demonstrate that it is neither biology alone nor environment alone that determines the final product of a personality.[26] When it comes to the nature versus nurture question, the answer almost always includes both.

As we saw in the previous chapter, you choose neither your nature nor your nurture, much less their entangled interaction. You inherit a genetic blueprint and are born into a world over which you have no choice throughout your most formative years. This is the reason people come to the table with quite different ways of seeing the world, dissimilar personalities, and varied capacities for decision making. These are not choices; these are the dealt hands of cards. The point of the previous chapter was to highlight the difficulty of assigning culpability under these circumstances. The point of this chapter is to highlight the fact that the machinery that makes us who we are is not simple, and that science is not perched on the verge of understanding how to build minds from pieces and parts. Without a doubt, minds and biology are connected—but not in a manner that we'll have any hope of understanding with a purely reductionist approach.

Reductionism is misleading for two reasons. First, as we have just seen, the unfathomable complexity of gene–environment interactions puts us a long way from understanding how any individual—with her lifetime of experiences, conversations, abuses, joys, ingested foods, recreational drugs, prescribed medications, pesticides, educational experience, and so on—will develop. It's simply too complex and will presumably remain so.

Second, even while it's true that we are tied to our molecules and proteins and neurons—as strokes and hormones and drugs and microorganisms indisputably tell us—it does not logically follow that humans are best described only as pieces and parts. The extreme reductionist idea that we are *no more than* the cells of which we are composed is a nonstarter for anyone trying to understand human behavior. Just because a system is made of pieces and parts, and just because those pieces and parts are critical to the working of the system, that does not mean that the pieces and parts are the correct level of description.

So why did reductionism catch on in the first place? To understand this, we need only to examine its historical roots. Over recent centuries, thinking men and women watched the growth of deterministic science around them in the form of the deterministic equations of Galileo, Newton, and others. These scientists pulled springs and rolled balls and dropped weights, and increasingly they were able to predict what the objects would do with simple equations. By the nineteenth century, Pierre-Simon Laplace had proposed that if one could know the position of every particle in the universe, then one could compute forward to know the entire future (and crank the equations in the other direction to know everything past). This historical success story is the heart of reductionism, which essentially proposes that everything big can be understood by discerning smaller and smaller pieces of it. In this viewpoint, the arrows of understanding all point to the smaller levels: humans can be understood in terms of biology, biology in the language of chemistry, and chemistry in the equations of atomic physics. Reductionism has been the engine of science since before the Renaissance.

But reductionism is not the right viewpoint for everything, and it certainly won't explain the relationship between the brain and the mind. This is because of a feature known as *emergence*.[27] When you put together large numbers of pieces and parts, the whole can become something greater than the sum. None of the individual metal hunks of an airplane have the property of *flight*, but when they are attached together in the right way, the result takes to the air. A thin metal bar won't do you much good if you're trying to control a jaguar, but several of them in parallel have the property of *containment*. The concept of emergent properties means that something new can be introduced that is not inherent in any of the parts.

As another example, imagine you were an urban highway planner and you needed to understand your city's traffic flow: where the cars tend to bunch up, where people speed, and where the most dangerous attempts at passing occur. It won't take you long to realize that an understanding of these issues will require some model of the psychology of the drivers themselves. You would lose your job if you proposed to study the length of the screws and the combustion efficiency of the spark plugs in the engines. Those are the wrong levels of description for understanding traffic jams.

This is not to say that the small pieces don't matter; they *do* matter. As we saw with brains, adding narcotics, changing neurotransmitter levels, or mutating genes can radically alter the essence of a person. Similarly, if you modify screws and spark plugs, the engines work differently, cars might speed or slow, and other cars might crash into them. So the conclusion is clear: while traffic flow depends on the integrity of the parts, it is not in any meaningful way *equivalent* to the parts. If you want to know why the television show *The Simpsons* is funny, you won't get far by studying the transistors and capacitors in the back of your plasma-screen television. You might be able to elucidate the electronic parts in great detail and probably learn a thing or two about electricity, but that won't get you any closer to understanding hilarity. Watching *The Simpsons* depends entirely on the integrity of the transistors, but the parts are not themselves funny. Similarly, while minds depend on the integrity of neurons, neurons are not themselves thinking.

And this forces a reconsideration of how to build a scientific account of the brain. If we were to work out a complete physics of neurons and their chemicals, would that elucidate the mind? Probably not. The brain presumably does not break the laws of physics, but that does not mean that equations describing detailed biochemical interactions will amount to the correct level of description. As the complexity theorist Stuart Kauffman puts it, "A couple in love walking along the banks of the Seine are, in real fact, a couple in love walking along the banks of the Seine, not mere particles in motion."

A meaningful theory of human biology cannot be reduced to chemistry and physics, but instead must be understood in its own vocabulary of evolution, competition, reward, desire, reputation, avarice, friendship, trust, hunger, and so on—in the same way that traffic flow will be understood not in the vocabulary of screws and spark plugs, but instead in terms of speed limits, rush hours, road rage, and people wanting to get home to their families as soon as possible when their workday is over.

There's another reason why the neural pieces and parts won't be sufficient for a full understanding of human experience: your brain is not the only biological player in the game of determining who you are. The brain is tied in constant two-way communication with the endocrine and immune systems, which can be thought of as the "greater nervous system." The greater nervous system is, in turn, inseparable from the chemical environments that influence its development—including nutrition, lead paint, air pollutants, and so on. And you are part of a complex social network that changes your biology with every interaction, and which your actions can change in return. This makes the borders interesting to contemplate: how should we define *you*? Where do you begin and where do you end? The only solution is to think about the brain as the densest concentration of *you*ness. It's the peak of the mountain, but not the whole mountain. When we talk about "the brain" and behavior, this is a shorthand label for something that includes contributions from a much broader sociobiological system.* The brain is not so much the seat of the mind as the hub of the mind.

So let's summarize where we are. Following a one-way street in the direction of the very small is the mistake that reductionists make, and it is the trap we want to avoid. Whenever you see a shorthand statement such as "you are your brain," don't understand it to mean that neuroscience will understand brains only as massive constellations of atoms or as vast jungles of neurons. Instead, the future of understanding the mind lies in deciphering the patterns

*In *Lifelines*, biologist Steven Rose points out that "reductionist ideology not only hinders biologists from thinking adequately about the phenomena we wish to understand: it has two important social consequences: it serves to relocate social problems to the individual . . . rather than exploring the societal roots and determinants of a phenomenon; and second, it diverts attention and funding from the social to the molecular."

of activity that live *on top of* the wetware, patterns that are directed both by internal machinations and by interactions from the surrounding world. Laboratories all over the world are working to figure out how to understand the relationship between physical matter and subjective experience, but it's far from a solved problem.

* * *

In the early 1950s, the philosopher Hans Reichenbach stated that humanity was poised before a complete, scientific, objective account of the world—a "scientific philosophy."[28] That was sixty years ago. Have we arrived? Not yet, anyway.

In fact, we're a long way off. For some people, the game is to act as though science is just on the brink of figuring everything out. Indeed, there is great pressure on scientists—applied from granting agencies and popular media alike—to pretend as though the major problems are about to be solved at any moment. But the truth is that we face a field of question marks, and this field stretches to the vanishing point.

This suggests an entreaty for openness while exploring these issues. As one example, the field of quantum mechanics includes the concept of *observation*: when an observer measures the location of a photon, that collapses the state of the particle to a particular position, while a moment ago it was in an infinity of possible states. What is it about *observation*? Do human minds interact with the stuff of the universe?[29] This is a totally unsolved issue in science, and one that will provide a critical meeting ground between physics and neuroscience. Most scientists currently approach the two fields as separate, and the sad truth is that researchers who try to look more deeply into the connections between them often end up marginalized. Many scientists will make fun of the pursuit by saying something like "Quantum mechanics is mysterious, and consciousness is mysterious; therefore, they must be the same thing." This dismissiveness is bad for the field. To be clear, I'm not asserting there *is* a connection between quantum mechanics and consciousness. I am saying there *could* be a connection, and that a premature dismissal is not in the spirit of scientific inquiry and progress. When people assert that brain function can be completely explained by classical physics, it is important to recognize that this is simply an assertion—it's difficult to know in any age of science what pieces of the puzzle we're missing.

As an example, I'll mention what I'll call the "radio theory" of brains. Imagine that you are a Kalahari Bushman and that you stumble upon a transistor radio in the sand. You might pick it up, twiddle the knobs, and suddenly, to your surprise, hear voices streaming out of this strange little box. If you're curious and scientifically minded, you might try to understand what is going on. You might pry off the back cover to discover a little nest of wires. Now let's say you begin a careful, scientific study of what causes the voices. You notice that each time you pull out the green wire, the voices stop. When you put the wire back on its contact, the voices begin again. The same goes for the red wire. Yanking out the black wire causes the voices to get garbled, and removing the yellow wire reduces the volume to a whisper. You step carefully through all the combinations, and you come to a clear conclusion: the voices depend entirely on the integrity of the circuitry. Change the circuitry and you damage the voices.

Proud of your new discoveries, you devote your life to developing a science of the way in which certain configurations of wires create the existence of magical voices. At some point, a young person asks you *how* some simple loops of electrical signals can engender music and conversations, and you admit that you don't know—but you insist that your science is about to crack that problem at any moment.

Your conclusions are limited by the fact that you know absolutely nothing about radio waves and, more generally, electromagnetic radiation. The fact that there are structures in distant cities called radio towers—which send signals by perturbing invisible waves that travel at the speed of light—is so foreign to you that you could not even dream it up. You can't taste radio waves, you can't see them, you can't smell them, and you don't yet have any pressing reason to be creative enough to fantasize about them. And if you *did* dream of invisible radio waves that carry voices, who could you convince of your hypothesis? You have no technology to demonstrate the existence of the waves, and everyone justifiably points out that the onus is on you to convince them.

So you would become a radio materialist. You would conclude that somehow the right configuration of wires engenders classical music and intelligent conversation. You would not realize that you're missing an enormous piece of the puzzle.

I'm not asserting that the brain is like a radio—that is, that we're receptacles picking up signals from elsewhere, and that our neural circuitry needs to be in place to do so—but I *am* pointing out that it *could* be true. There is nothing in our current science that rules this out. Knowing as little as we do at this point in history, we must retain concepts like this in the large filing cabinet of ideas that we cannot yet rule in favor of or against. So even though few working scientists will design experiments around eccentric hypotheses, ideas always need to be proposed and nurtured as possibilities until evidence weighs in one way or another.

Scientists often talk of parsimony (as in "the simplest explanation is probably correct," also known as Occam's razor), but we should not get seduced by the apparent elegance of argument from parsimony; this line of reasoning has failed in the past at least as many times as it has succeeded. For example, it is more parsimonious to assume that the sun goes around the Earth, that atoms at the smallest scale operate in accordance with the same rules that objects at larger scales follow, and that we perceive what is really out there. All of these positions were long defended by argument from parsimony, and they were all wrong. In my view, the argument from parsimony is really no argument at all—it typically functions only to shut down more interesting discussion. If history is any guide, it's never a good idea to assume that a scientific problem is cornered.

At this moment in history, the majority of the neuroscience

community subscribes to materialism and reductionism, enlisting the model that we are understandable as a collection of cells, blood vessels, hormones, proteins, and fluids—all following the basic laws of chemistry and physics. Each day neuroscientists go into the laboratory and work under the assumption that understanding enough of the pieces and parts will give an understanding of the whole. This break-it-down-to-the-smallest-bits approach is the same successful method that science has employed in physics, chemistry, and the reverse-engineering of electronic devices.

But we don't have any real guarantee that this approach will work in neuroscience. The brain, with its private, subjective experience, is unlike any of the problems we have tackled so far. Any neuroscientist who tells you we have the problem cornered with a reductionist approach doesn't understand the complexity of the problem. Keep in mind that every single generation before us has worked under the assumption that they possessed all the major tools for understanding the universe, and they were all wrong, without exception. Just imagine trying to construct a theory of rainbows before understanding optics, or trying to understand lightning before knowledge of electricity, or addressing Parkinson's disease before the discovery of neurotransmitters. Does it seem reasonable that we are the first ones lucky enough to be born in the perfect generation, the one in which the assumption of a comprehensive science is finally true? Or does it seem more likely that in one hundred years people will look back on us and wonder what it was like to to be ignorant of what they know? Like the blind people in Chapter 4, we do not experience a gaping hole of blackness where we are lacking information—instead, we do not appreciate that anything is missing.[30]

I'm not saying that materialism is incorrect, or even that I'm hoping it's incorrect. After all, even a materialist universe would be mind-blowingly amazing. Imagine for a moment that we are nothing but the product of billions of years of molecules coming together and ratcheting up through natural selection, that we are composed only of highways of fluids and chemicals sliding along roadways within billions of dancing cells, that trillions of synaptic conversations hum in parallel, that this vast egglike fabric of micronthin circuitry runs algorithms undreamt of in modern science, and that these neural programs give rise to our decision making, loves, desires, fears, and aspirations. To me, that understanding would be a numinous experience, better than anything ever proposed in anyone's holy text. Whatever else exists beyond the limits of science is an open question for future generations; but even if strict materialism turned out to be it, it would be enough.

Arthur C. Clarke was fond of pointing out that any sufficiently advanced technology is indistinguishable from magic. I don't view the dethronement from the center of ourselves as depressing; I view it as magic. We've seen in this book that everything contained in the biological bags of fluid we call *us* is already so far beyond our intuition, beyond our capacity to think about such vast scales of interaction, beyond our introspection that this fairly qualifies as "something beyond us." The complexity of the system we *are* is so vast as to be indistinguishable from Clarke's magical technology. As the quip goes: If our brains were simple enough to be understood, we wouldn't be smart enough to understand them.

In the same way that the cosmos is larger than we ever imagined, we ourselves are something greater than we had intuited by introspection. We're now getting the first glimpses of the vastness of inner space. This internal, hidden, intimate cosmos commands its own goals, imperatives, and logic. The brain is an organ that feels alien and outlandish to us, and yet its detailed wiring patterns sculpt the landscape of our inner lives. What a perplexing masterpiece the brain is, and how lucky we are to be in a generation that has the technology and the will to turn our attention to it. It is the most wondrous thing we have discovered in the universe, and it is us.

Appendix

Dramatis Personae

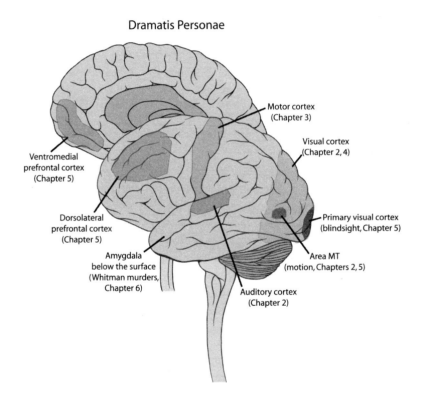

Motor cortex
(Chapter 3)

Visual cortex
(Chapter 2, 4)

Ventromedial
prefrontal cortex
(Chapter 5)

Primary visual cortex
(blindsight, Chapter 5)

Dorsolateral
prefrontal cortex
(Chapter 5)

Area MT
(motion, Chapters 2, 5)

Amygdala
below the surface
(Whitman murders,
Chapter 6)

Auditory cortex
(Chapter 2)

Acknowledgments

Many people have inspired me in the writing of this book. Some had atoms that went their separate ways before my atoms came together—I may have inherited some of their atoms, but, more importantly, I was fortunate enough to inherit the ideas they left behind as messages in a bottle. I have also been fortunate to share existence contemporaneously with a network of tremendously smart people, which began with my parents, Arthur and Cirel, and continued with my graduate thesis adviser, Read Montague, and was sustained by mentors such as Terry Sejnowski and Francis Crick at the Salk Institute. I enjoy daily inspiration with colleagues, students, and friends such as Jonathan Downar, Brett Mensh, Chess Stetson, Don Vaughn, Abdul Kudrath, and Brian Rosenthal, to name a few. I thank Dan Frank and Nick Davies for their expert editorial feedback, and Tina Borja and all the students in my lab for line-by-line readings; these students include Tommy Sprague, Steffie Tomson, Ben Bumann, Brent Parsons, Mingbo Cai, and Daisy Thomson-Lake. I thank Jonathan D. Cohen for a seminar he gave which shaped some of my thinking in Chapter 5. Thank you to Shaunagh Darling Robertson for proposing the title *Incognito*. I am grateful to launch my books from the solid bedrock of the Wylie Agency, which includes the gifted Andrew Wylie, the exceptional Sarah Chalfant, and all their skilled coworkers. I am deeply grateful to my first agent, Jane Gelfman, for believing in me and in this book from the beginning. I thank Jamie Byng for his boundless enthusiasm and deep support. Finally, my gratitude goes to my wife Sarah for her love, humor and encouragement. The other day I saw a sign that simply read HAPPINESS—and I realized that the thought of Sarah was my instant mental headline. Deep in the canopies of my brain forest, happiness and Sarah have become synaptically synonymous, and for her presence in my life I am grateful.

*   *   *

Throughout the book you will often find the narrator's term we instead of I. This is for three reasons. First, as with any book that synthesizes large bodies of knowledge, I collaborate with thousands of scientists and historians over the course of centuries. Second, the reading of a book should be an active collaboration between reader and writer. Third, our brains are composed of vast, complex, and shifting collections of subparts, most of which we have no access to; this book was written over the course of a few years by several different people, all of whom were named David Eagleman, but who were somewhat different with each passing hour.

Notes

Works listed in full in the Bibliography are referred to only by short title here.

Chapter 1. There's Someone In My Head, But It's Not Me

1 Music: "Tremendous Magic," *Time* December 4, 1950.

2 Something I've always found inspiring: the year Galileo died—1642—Isaac Newton was born into the world and completed Galileo's job by describing the equations underlying the planetary orbits around the sun.

3 Aquinas, *Summa theologiae*.

4 Specifically, Leibniz envisioned a machine that would use marbles (representing binary numbers) that would be guided by what we now recognize as cousins to punch cards. Although Charles Babbage and Ada Lovelace are generally credited with working out the concepts of software separation, the modern computer is essentially no different than what Leibniz envisaged: "This [binary] calculus could be implemented by a machine (without wheels) in the following manner, easily to be sure and without effort. A container shall be provided with holes in such a way that they can be opened and closed. They are to be open at those places that correspond to a 1 and remain closed at those that correspond to a 0. Through the opened gates small cubes or marbles are to fall into tracks, through the others nothing. It [the gate array] is to be shifted from column to column as required." See Leibniz, *De*

Progressione Dyadica. Thanks to George Dyson for this discovery in the literature.

5 Leibniz, *New Essays on Human Understanding*, published 1765. By "insensible corpuscles," Leibniz is referring to the belief shared by Newton, Boyle, Locke, and others that material objects are made of tiny insensible corpuscles, which give rise to the sense qualities of the objects.

6 Herbart, *Psychology as a Science*.

7 Michael Heidelberger, *Nature from Within*.

8 Johannes Müller, *Handbuch der Physiologie des Menschen, dritte verbesserte Auflage*, 2 vols (Coblenz: Hölscher, 1837–1840).

9 Cattell, "The time taken up," 220–242.

10 Cattell, "The psychological laboratory," 37–51.

11 See http://www.iep.utm.edu/f/freud.htm.

12 Freud and Breuer, *Studien über Hysterie*.

Chapter 2. The Testimony of the Senses

1 Eagleman, "Visual illusions."

2 Sherrington, *Man on His Nature*. See also Sheets-Johnstone, "Consciousness: a natural history."

3 MacLeod and Fine, "Vision after early blindness."

4 Eagleman, "Visual illusions."

5 See eagleman.com/incognito for interactive demonstrations of how little we perceive of the world. For excellent reviews on change blindness, see Rensink, O'Regan, and Clark, "To see or not to see"; Simons, "Current approaches to change blindness"; and Blackmore, Brelstaff, Nelson, and Troscianko, "Is the richness of our visual world an illusion?"

6 Levin and Simons, "Failure to detect changes to attended objects."

7 Simons and Levin, "Failure to detect changes to people."

8 Macknik, King, Randi, et. al., "Attention and awareness in stage magic."

9 The concept of a 2.5-D sketch was introduced by the late neuroscientist David Marr. He originally proposed this as an intermediate stage on the visual system's journey to developing a full 3-D model, but it has since become clear that the full 3-D model never comes to fruition in real brains, and is not needed to get by in the world. See Marr, *Vision*.

10 O'Regan, "Solving the real mysteries of visual perception," and Edelman, *Representation and Recognition in Vision*. Note that one group recognized the problem early on, in 1978, but it took many years to become more widely recognized: "The primary function of perception is to keep our internal framework in good registration with that vast external memory, the external environment itself," noted Reitman, Nado, and Wilcox in "Machine perception," 72.

11 Yarbus, "Eye movements."

12 This phenomenon is known as binocular rivalry. For reviews, see Blake and Logothetis, "Visual competition" and Tong, Meng, and Blake, "Neural bases."

13 The hole of missing photoreceptors occurs because the optic nerve passes through this location in the retina, leaving no room for the light-sensing cells. Chance, "Ophthalmology," and Eagleman, "Visual illusions."

14 Helmholtz, *Handbuch*.

15 Ramachandran, "Perception of shape."

16 Kersten, Knill, Mamassian, and Bülthoff, "Illusory motion."

17 Mather, Verstraten, and Anstis, *The Motion Aftereffect*, and Eagleman, "Visual illusions."

18 Dennett, *Consciousness Explained*.

19 Baker, Hess, and Zihl, "Residual motion"; Zihl, von Cramon, and Mai, "Selective disturbance"; and Zihl, von Cramon, Mai, and Schmid, "Disturbance of movement vision."

20 McBeath, Shaffer, and Kaiser, "How baseball outfielders."

21 It turns out that fighter pilots use this same algorithm during pursuit tasks, as do fish and hoverflies. Pilots: O'Hare, "Introduction"; fish: Lanchester and Mark, "Pursuit and prediction"; and hoverflies: Collett and Land, "Visual control."

22 Kurson, *Crashing Through*.

23 It should be noted that some blind people can convert their felt world to two- or three-dimensional drawings. However, it is presumably the case that drawing the converging lines of a hallway is a cognitive exercise for them, different from the way that sighted people have the immediate sensory experience.

24 Noë, *Action in Perception.*

25 P. Bach-y-Rita, "Tactile sensory substitution studies."

26 Bach-y-Rita, Collins, Saunders, White, and Scadden, "Vision substitution."

27 For an overview and synthesis of these studies, see Eagleman, *Live-Wired*. Nowadays it is popular for the tactile display to come from an electrode grid placed directly on the tongue. See Bach-y-Rita, Kaczmarek, Tyler, and Garcia-Lara, "Form perception."

28 Eagleman, *Live-Wired.*

29 C. Lenay, O. Gapenne, S. Hanneton, C. Marque, and C. Genouel, "Sensory substitution: Limits and perspectives," in *Touching for Knowing, Cognitive Psychology of Haptic Manual Perception* (Amsterdam: John Benjamins, 2003), 275–92, and Eagleman, *Live-Wired.*

30 The BrainPort is made by Wicab, Inc, a company founded by plasticity pioneer Paul Bach-y-Rita.

31 Bach-y-Rita, Collins, Saunders, White, and Scadden, "Vision substitution"; Bach-y-Rita, "Tactile sensory substitution studies"; Bach-y-Rita, Kaczmarek, Tyler, and Garcia-Lara, "Form perception"; M. Ptito, S. Moesgaard, A. Gjedde, and R. Kupers, "Cross-modal plasticity revealed by electrotactile stimulation of the tongue in the congenitally blind," *Brain* 128 (2005), 606–14; and Bach-y-Rita, "Emerging concepts of brain function," *Journal of Integrative Neuroscience* 4 (2005), 183–205.

32 Yancey Hall. "Soldiers may get 'sight' on tips of their tongues," *National Geographic News*, May 1, 2006.

33 B. Levy, "The blind climber who 'sees' with his tongue," *Discover*, June 23, 2008.

34 Hawkins, *On Intelligence*, and Eagleman, *Live-Wired.*

35 Gerald H. Jacobs, Gary A. Williams, Hugh Cahill, and Jeremy Nathans, "Emergence of novel color vision in mice engineered to express a human cone photopigment," *Science* 23 (2007): vol. 315. no. 5819, 1723–25. For a detracting opinion about the interpretation of results, see Walter Makous, "Comment on 'Emergence of novel color vision in mice engineered to express a human cone photopigment,'" *Science*

(2007): vol. 318. no. 5848, 196, in which he argues that it is impossible to conclude much of anything about the internal experience of the mice, a precondition for claiming that they experienced color vision as opposed to different levels of light and dark. Whatever the internal experience of the mice, it is clear that their brains have integrated the information from the new photopigments and can now discriminate features that they could not before. Importantly, this technique is now possible in rhesus monkeys, a method that should open the door to asking the correct, detailed perceptual questions.

36 Jameson, "Tetrachromatic color vision."

37 Llinas, *I of the Vortex.*

38 Brown, "The intrinsic factors." Although Brown was well known in the 1920s for his pioneering neurophysiology experiments, he became even better known in the 1930s for his world-famous mountaineering expeditions and discoveries of new routes to the summit of Mont Blanc.

39 Bell, "Levels and loops."

40 McGurk and MacDonald, "Hearing lips," and Schwartz, Robert-Ribes, and Escudier, "Ten years after Summerfield."

41 Shams, Kamitani, and Shimojo, "Illusions."

42 Gebhard and Mowbray, "On discriminating"; Shipley, "Auditory flutter-driving"; and Welch, Duttonhurt, and Warren, "Contributions."

43 Tresilian, "Visually timed action"; Lacquaniti, Carrozzo, and Borghese, "Planning and control of limb impedance"; Zago, et. al., "Internal models"; McIntyre, Zago, Berthoz, and Lacquaniti, "Does the brain model Newton's laws?"; Mehta and Schaal, "Forward models"; Kawato, "Internal models"; Wolpert, Ghahramani, and Jordan, "An internal model"; and Eagleman, "Time perception is distorted during visual slow motion," Society for Neuroscience, abstract, 2004.

44 MacKay, "Towards an information-flow model"; Kenneth Craik, *The Nature of Explanation* (Cambridge, UK: Cambridge University Press, 1943); Grush, "The emulation theory". Also see Kawato, Furukawa, and Suzuki, "A hierarchical neural-network model"; Jordan and Jacobs, "Heirarchical mixtures of experts"; Miall and Wolpert, "Forward models"; and Wolpert and Flanagan, "Motor prediction."

45 Grossberg, "How does a brain . . . ?"; Mumford, "On the computational

architecture"; Ullman, "Sequence seeking"; and Rao, "An optimal estimation approach."

46 MacKay, "The epistemological problem."

47 See Blakemore, Wolpert, and Frith, "Why can't you tickle yourself?" for more about tickling. More generally, violations of sensory expectations can inform a brain about responsibility—that is, did I cause the action or did someone else? Schizophrenic hallucinations may arise from a failure of matching expectations about one's own motor acts to their resulting sensory signals. Failure to distinguish one's own actions from those of independent agents means that the patient attributes his internal voices to someone else. For more on this idea, see Frith and Dolan, "Brain mechanisms."

48 Symonds and MacKenzie, "Bilateral loss of vision."

49 Eagleman and Sejnowski, "Motion integration," and Eagleman, "Human time perception."

50 Eagleman and Pariyadath, "Is subjective duration . . . ?"

Chapter 3. Mind: The Gap

1 Macuga, et al., "Changing lanes."

2 Schacter, "Implicit memory."

3 Ebbinghaus, *Memory: A Contribution to Experimental Psychology.*

4 Horsey, *The Art of Chicken Sexing*; Biederman and Shiffrar, "Sexing day-old chicks"; Brandom, "Insights and blindspots of reliabilism"; and Harnad, "Experimental analysis."

5 Allan, "Learning perceptual skills."

6 Cohen, Eichenbaum, Deacedo, and Corkin, "Different memory systems," and Brooks and Baddeley, "What can amnesic patients learn?"

7 As another example of tying things together at an unconscious level, subjects were given a carbonated drink, and then their chairs were rocked back and forth to induce motion sickness. As a result, the subjects established an aversion to the carbonated drink, even though they well knew (consciously) that the drink had nothing to do with the nauseating motion. See Arwas, Rolnick, and Lubow, "Conditioned taste aversion."

8 Greenwald, McGhee, and Schwartz, "Measuring individual differences."

9 The implicit association test can be taken online: https://implicit.harvard.edu/implicit/demo/selectatest.html.

10 Wojnowicz, Ferguson, Dale, and Spivey, "The self-organization of explicit attitudes." See also Freeman, Ambady, Rule, and Johnson, "Will a category cue attract you?"

11 Jones, Pelham, Carvallo, and Mirenberg, "How do I love thee?"

12 Ibid.

13 Pelham, Mirenberg, and Jones, "Why Susie sells," and Pelham, Carvallo, and Jones, "Implicit egotism."

14 Abel, "Influence of names."

15 Jacoby and Witherspoon, "Remembering without awareness."

16 Tulving, Schacter, and Stark, "Priming effects." These effects hold even if I distract you so much that we're certain you cannot explicitly remember what the words were; you're still just as good at the word completion. See Graf and Schacter, "Selective effects."

17 The idea of priming has a rich history in literature and entertainment. In *The Subliminal Man*, by J. G. Ballard (1963), a character named Hathaway is the only one who suspects that the dozens of gigantic blank signs towering over the roads are really subliminal advertising machines, encouraging people to take on more jobs and buy more products. A more droll incarnation of Subliminal Man can be found in comedian Kevin Nealon's *Saturday Night Live* character, who says, during a talk show interview, "I've always liked watching this show (nauseating). It's fun to be a guest on this show (torture). It's kind of like a second home to me (*Titanic*)."

18 Graf and Schacter, "Implicit and explicit memory."

19 See Tom, Nelson, Srzentic, and King, "Mere exposure." For a more basic approach to demonstrating that the brain can absorb what it has seen even without attention to it, see Gutnisky, Hansen, Iliescu, and Dragoi, "Attention alters visual plasticity."

20 Ironically, no one is quite sure who said this first. The quotation has been variously attributed to Mae West, P. T. Barnum, George M. Cohan, Will Rogers, and W. C. Fields, among others.

21 Hasher, Goldstein, and Toppino, "Frequency and the conference of referential validity."

22 Begg, Anas, and Farinacci, "Dissociation of processes in belief."

23 Cleeremans, *Mechanisms of Implicit Learning.*

24 Bechara, Damasio, Tranel, and Damasio, "Deciding advantageously."

25 Damasio, "The somatic marker hypothesis"; Damasio, *Descartes' Error*; and Damasio, *The Feeling of What Happens.*

26 Eagleman, *Live-Wired.*

27 Montague, *Your Brain Is (Almost) Perfect.*

28 If you watch athletes closely, you will notice they often employ physical rituals to get themselves into the zone—for example, dribbling the ball exactly three times, cranking their neck to the left, and then shooting. By providing predictability, these rituals ease them into a less conscious state. To the same end, repetitive and predictable rituals are routinely used in religious services—for example, rote prayers, rosary counting, and chants all help to relax the buzzing of the conscious mind.

Chapter 4. The Kinds of Thoughts That Are Thinkable

1 Blaise Pascal, *Pensées*, 1670.

2 All of these signals (radio, microwave, X-ray, gamma ray, cell phone transmissions, television broadcasts, and so on) are exactly the same thing as the stuff coming out of the front of your flashlight—just of a different wavelength. Some readers already knew this; for those who didn't, the sheer amazingness of this simple scientific fact urges its inclusion.

3 Jakob von Uexküll introduced the idea of the umwelt in 1909 and explored it through the 1940s. It was then lost for decades until rediscovery and revivification by the semiotician Thomas A. Sebeok in 1979; Jakob von Uexküll, "A stroll through the worlds of animals and men." See also Giorgio Agamben, Chapter 10, "Umwelt", in his book *The Open: Man and Animal*, trans. Kevin Attell (Palo Alto: Stanford University Press, 2004); originally published in Italian in 2002 as *L'aperto: l'uomo e l'animale.*

4 K. A. Jameson, S. Highnote, and L. Wasserman, "Richer color

experience in observers with multiple photopigment opsin genes," *Psychonomic Bulletin & Review*, 8, no.2 (2001): 244–61; and Jameson, "Tetrachromatic color vision."

5 For more about synesthesia, see Cytowic and Eagleman, *Wednesday Is Indigo Blue*.

6 Think you might have synesthesia? Take the free online tests at www.synesthete.org. See Eagleman, et al., "A standardized test battery for the study of synesthesia."

7 Our laboratory has turned to the details of synesthesia—from behavior to neuroimaging to genetics—to use it as an inroad into understanding how slight differences in the brain can lead to large differences in the perception of reality. See www.synesthete.org.

8 In other words, the forms have a location in mental space that can be pointed to. If you're not a spatial sequence synesthete, imagine your car parked in the space in front of you. Although you do not physically see it there like a hallucination, you will have no trouble pointing to the front wheel, the driver's side window, the rear bumper, and so on. The car has three-dimensional coordinates in your mental space. So it goes with automatically triggered number forms. Unlike hallucinations, they do not overlie the outside visual world; they instead live in a mental space. In fact, even blind subjects can experience number form synesthesia; see Wheeler and Cutsforth, "The number forms of a blind subject." For a larger discussion of spatial sequence synesthesia, see Eagleman, "The objectification of overlearned sequences," and Cytowic and Eagleman, *Wednesday Is Indigo Blue*.

9 Eagleman, "The objectification of overlearned sequences."

10 An interesting speculation is that all brains are synesthetic—but the majority of us remain unconscious of the sensory fusions going on in our brains under the surface of awareness. In fact, everyone seems to possess implicit number lines for sequences. When asked, we might agree that the number line for integers increases as one goes from left to right. Spatial sequence synesthetes differ in that they experience sequences explicitly in three dimensions as automatic, consistent, and concrete configurations. See Eagleman, "The objectification of over-

learned sequences", and Cytowic and Eagleman, *Wednesday Is Indigo Blue*.

11 Nagel, *The View from Nowhere*.

12 See Cosmides and Tooby, *Cognitive Adaptations*, for an overview, and Steven Pinker's *The Blank Slate* for an excellent in-depth read.

13 Johnson and Morton, "CONSPEC and CONLERN."

14 Meltzoff, "Understanding the intentions of others."

15 Pinker, *The Blank Slate*.

16 Wason and Shapiro, "Reasoning," and Wason, "Natural and contrived experience"

17 Cosmides and Tooby, *Cognitive Adaptions*.

18 Barkow, Cosmides, and Tooby, *The Adapted Mind*

19 Cosmides and Tooby, "Evolutionary psychology: A primer," 1997; http://www.psych.ucsb.edu/research/cep/primer.html

20 James, *The Principles of Psychology*.

21 Tooby and Cosmides, *Evolutionary Psychology: Foundational Papers* (Cambridge, MA: MIT Press, 2000).

22 Singh, "Adaptive significance" and "Is thin really beautiful," and Yu and Shepard, "Is beauty in the eye?"

23 More generally, women with waists thinner than this range are seen as more aggressive and ambitious, while those with thicker waists are viewed as kind and faithful. On another note, one reader correctly pointed out that I should acknowledge the stresses, the losses and the heartbreak of being attractive, especially for women. Nancy Etcoff reviews the literature thoroughly in her book *Survival of the Prettiest*, explaining that although beauty may get a woman the job, it often prevents her from getting the promotion.

24 Ramachandran, "Why do gentlemen prefer blondes?" After publishing his article, Ramachandran subsequently claimed that he meant his publication as a hoax, a satire on sociobiological theories of human mate selection (see his *Phantoms in the Brain*). Nonetheless, it's not really a hoax in the traditional sense, especially as (1) it's not necessarily incorrect, and (2) Ramachandran is not willing to distance himself from it entirely (he gives it a "less than ten percent chance of being correct", not a zero chance). As he put it in his original *Medical Hypotheses*

publication: "Although originally intended as a satire on *ad hoc* socio-biological theories of human mate-selection, I soon came to realize that this idea is at least as viable as many other theories of mate choice that are currently in vogue."

25 Penton-Voak, et al., "Female preference for male faces changes cyclically."

26 Vaughn and Eagleman, "Faces briefly glimpsed."

27 Friedman, McCarthy, Förster, and Denzler, "Automatic effects." It may even be the case that other concepts related to alcohol (such as sociability) may also be activated by priming to alcohol-related words—such that merely seeing (not drinking) a glass of wine could lead to easier conversation and more eye contact. A more speculative and challenging possibility is that seeing advertisements for alcohol along highway billboards could lead to diminished driving performance.

28 Concealed ovulation (as well as internal fertilization, as opposed to the laying of external eggs) may have come about as a mechanism that encourages males to remain attentive to their female mates equally at all times, thereby diminishing the chances of desertion.

29 Roberts, Havlicek, and Flegr, "Female facial attractiveness increases."

30 Symmetry of ears, breasts, and fingers during ovulation: Manning, Scutt, Whitehouse, Leinster, and Walton, "Asymmetry," Scutt and Manning, "Symmetry"; for lighter skin tone, see Van den Berghe and Frost, "Skin color preference."

31 G. F. Miller, J. M. Tybur, and B. D. Jordan, "Ovulatory cycle effects on tip earnings by lap-dancers: Economic evidence for human estrus?" *Evolution and Human Behavior*, 28 (2007): 375–81.

32 Liberles and Buck, "A second class." Because humans also carry the genes for this family of receptors, it's the most promising road to sniff down when looking for a role for pheromones in humans.

33 Pearson, "Mouse data."

34 C. Wedekind, T. Seebeck, F. Bettens, and A. J. Paepke, "MHC-dependent mate preferences in humans." *Proceeding of the Royal Society of London Series B: Biological Sciences* 260, no. 1359 (1995): 245–49.

35 Varendi and Porter, "Breast odour."

36 Stern and McClintock, "Regulation of ovulation by human pheromones." While it is widely believed that women living together will synchronize their menstrual cycles, it appears that this is not true. Careful studies of the original reports (and subsequent large-scale studies) show that statistical fluctuations can give the *perception* of synchrony but are nothing but chance occurrence. See Zhengwei and Schank, "Women do not synchronize."

37 Moles, Kieffer, and D'Amato, "Deficit in attachment behavior."

38 Lim, et al., "Enhanced partner preference."

39 H. Walum, L. Westberg, S. Henningsson, J. M. Neiderhiser, D. Reiss, W. Igl, J. M. Ganiban, et al.,"Genetic variation in the vasopressin receptor 1a gene (AVPR1A) associates with pair-bonding behavior in humans." *PNAS* 105, no.37 (2008): 14153–56.

40 Winston, *Human Instinct.*

41 Fisher, *Anatomy of Love.*

Chapter 5. The Brain Is a Team of Rivals

1 See Marvin Minsky's 1986 book *Society of Mind.*

2 Diamond, *Guns, Germs, and Steel.*

3 For a concrete illustration of the advantages and shortcomings of a "society" architecture, consider the concept of subsumption architecture, pioneered by the roboticist Rodney Brooks (Brooks, "A robust layered"). The basic unit of organization in the subsumption architecture is a module. Each module specializes in some independent, low-level task, such as controlling a sensor or actuator. The modules operate independently, each doing its own task. Each module has an input and an output signal. When the input of a module exceeds a predetermined threshold, the output of the module is activated. Inputs come from sensors or other modules. Each module also accepts a suppression signal and an inhibition signal. A suppression signal overrides the normal input signal. An inhibition signal causes output to be completely inhibited. These signals allow behaviors to override each other so that the system can produce coherent behavior. To produce coherent behavior, the modules are organized into layers. Each layer might implement a behavior, such as *wander* or *follow a moving object.*

These layers are hierarchical: higher layers can inhibit the behavior of lower ones by inhibition or suppression. This gives each level its own rank of control. This architecture tightly couples perception and action, producing a highly reactive machine. But the downside is that all patterns of behavior in these systems are prewired. Subsumption agents are fast, but they depend entirely on the world to tell them what to do; they are purely reflexive. In part, subsumption agents have far-from-intelligent behavior because they lack an internal model of the world from which to make conclusions. Rodney Brooks claims this is an advantage: by lacking representation, the architecture avoids the time necessary to read, write, utilize, and maintain the world models. But somehow, human brains *do* put in the time, and have clever ways of doing it. I argue that human brains will be simulated only by moving beyond the assembly line of sequestered experts into the idea of a conflict-based democracy of mind, where multiple parties pitch in their votes on the same topics.

4 For example, this approach is used commonly in artificial neural networks: Jacobs, Jordan, Nowlan, and Hinton, "Adaptive mixtures."

5 Minsky, *Society of Mind*.

6 Ingle, "Two visual systems," discussed in a larger framework by Milner and Goodale, *The Visual Brain*.

7 For the importance of conflict in the brain, see Edelman, *Computing the Mind*. An optimal brain can be composed of conflicting agents; see Livnat and Pippenger, "An optimal brain"; Tversky and Shafir, "Choice under conflict"; Festinger, *Conflict, Decision, and Dissonance*. See also Cohen, "The vulcanization," and McClure et al., "Conflict monitoring."

8 Miller, "Personality," as cited in Livnat and Pippenger, "An optimal brain."

9 For a review of dual-process accounts, see Evans, "Dual-processing accounts."

10 See Table 1 of ibid.

11 Freud, *Beyond the Pleasure Principle* (1920). The ideas of his three-part model of the psyche were expanded three years later in his *Das Ich und das Es*, available in Freud, *The Standard Edition*.

12 See, for example: Mesulam, *Principles of Behavioral and Cognitive neurology*; Elliott, Dolan, and Frith, "Dissociable functions"; and Faw, "Pre-frontal executive committee." There are many subtleties of the neuroanatomy and debates within the field, but these details are not central to my argument and will therefore be relegated to these references.

13 Some authors have referred to these systems, dryly, as System 1 and System 2 processes (see, for example, Stanovich, *Who is rational?* or Kahneman and Frederick, "Representativeness revisited"). For our purposes, we use what we hope will be the most intuitive (if imperfect) use of emotional and rational systems. This choice is common in the field; see, for example, Cohen, "The vulcanization," and McClure, et al., "Conflict monitoring."

14 In this sense, emotional responses can be viewed as information processing—every bit as complex as a math problem but occupied with the internal world rather than the outside. The output of their processing—brain states and bodily responses—can provide a simple plan of action for the organism to follow: do this, don't do that.

15 Greene, et al., "The neural bases of cognitive conflict."

16 See Niedenthal, "Embodying emotion," and Haidt, "The new synthesis."

17 Frederick, Loewenstein, and O'Donoghue, "Time discounting."

18 McClure, Laibson, Loewenstein, and Cohen, "Separate neural systems." Specifically, when choosing longer-term rewards with higher return, the lateral prefrontal and posterior parietal cortices were more active.

19 R. J. Shiller, "Infectious exuberance," *Atlantic Monthly*, July/August 2008.

20 Freud, "The future of an illusion," in *The Standard Edition*.

21 Illinois *Daily Republican*, Belvidere, IL, January 2, 1920.

22 Arlie R. Slabaugh, *Christmas Tokens and Medals* (Chicago: printed by Author, 1966), ANA Library Catalogue No. RM85.C5S5.

23 James Surowiecki, "Bitter money and christmas clubs," *Forbes.com*, February 14, 2006.

24 Eagleman, "America on deadline."

25 Thomas C. Schelling, *Choice and Consequence* (Cambridge, MA

Harvard University Press, 1984); Ryan Spellecy, "Reviving Ulysses contracts," *Kennedy Institute of Ethics Journal* 13, no. 4 (2003): 373–92; Namita Puran, "Ulysses contracts: Bound to treatment or free to choose?" *York Scholar* 2 (2005): 42–51.

26 There is no guarantee that the ethics boards accurately guess at the mental life of the future patient; then again, Ulysses contracts always suffer from imperfect knowledge of the future.

27 This phrase is borrowed from my colleague Jonathan Downar, who put it as "If you can't rely on your own dorsolateral prefrontal cortex, borrow someone else's." As much as I love the original phrasing, I've simplified it for the present purposes.

28 For a detailed summary of decades of split-brain studies, see Tramo, et al., "Hemispheric Specialization." For a lay-audience summary, see Michael Gazzaniga, "The split-brain revisited."

29 Jaynes, *The Origin of Consciousness.*

30 See, for example, Rauch, Shin, and Phelps, "Neurocircuitry models." For an investigation of the relationship between fearful memories and the perception of time, see Stetson, Fiesta, and Eagleman, "Does time really . . . ?"

31 Here's another aspect to consider about memory and the ceaseless reinvention hypothesis: neuroscientists do not think of memory as one phenomenon but, instead, as a collection of many different subtypes. On the broadest scale, there is short-term and long-term memory. Short-term involves remembering a phone number long enough to dial it. Within the long-term category there is declarative memory (for example, what you ate for breakfast and what year you got married) and non-declarative memory (how to ride a bicycle); for an overview, see Eagleman and Montague, "Models of learning." These divisions have been introduced because patients can sometimes damage one subtype without damaging others—an observation that has led neuroscientists into a hope of categorizing memory into several silos. But it is likely that the final picture of memory won't divide so neatly into natural categories; instead, as per the theme of this chapter, different memory mechanisms will *overlap* in their domains. (See, for example, Poldrack and Packard, "Competition," for a review of separable "cognitive" and

"habit" memory systems that rely on the medial temporal lobe and basal ganglia, respectively.) Any circuit that contributes to memory, even a bit, will be strengthened and can make its contribution. If true, this will go some distance toward explaining an enduring mystery to young residents entering the neurology clinic: why do real patient cases only rarely match the textbook case descriptions? Textbooks assume neat categorization, while real brains ceaselessly reinvent overlapping strategies. As a result, real brains are robust—and they are also resistant to humancentric labeling.

32 For a review of different models of motion detection, see Clifford and Ibbotson, "Fundamental mechanisms."

33 There are many examples of this inclusion of multiple solutions in modern neuroscience. Take, for instance, the motion aftereffect, mentioned in Chapter 2. If you stare at a waterfall for a minute or so, then look away at something else—say, the rocks on the side—it will look as though the stationary rocks are moving upward. This illusion results from an adaptation of the system; essentially, the visual brain realizes that it is deriving little new information from all the downward motion, and it starts to adjust its internal parameters in the direction of canceling out the downwardness. As a result, something stationary now begins to look like it's moving upward. For decades, scientists debated whether the adaptation happens at the level of the retina, at the early stages of the visual system, or at later stages of the visual system. Years of careful experiments have finally resolved this debate by dissolving it: there is no single answer to the question, because it is ill-posed. There is adaptation at many different levels in the visual system (Mather, Pavan, Campana, and Casco, "The motion aftereffect"). Some areas adapt quickly, some slowly, others at speeds in between. This strategy allows some parts of the brain to sensitively follow changes in the incoming data stream, while others will not change their stubborn ways without lasting evidence. Returning to the issue of memory discussed above, it is also theorized that Mother Nature has found several methods to store memories at several different time scales, and it is the interaction of all these time scales that makes older memories more stable than young memories. The fact that older memories are

more stable is known as Ribot's law. For more on the idea of exploiting different time scales of plasticity, see Fusi, Drew, and Abbott, "Cascade models."

34 In a wider biological context, the team-of-rivals framework accords well with the idea that the brain is a Darwinian system, one in which stimuli from the outside world happen to resonate with certain random patterns of neural circuitry, and not with others. Those circuits that happen to respond to stimuli in the outside world are strengthened, and other random circuits continue to drift around until they find something to resonate with. If they never find anything to "excite" them, they die off. To phrase it from the opposite direction, stimuli in the outside world "pick out" circuits in the brain: they happen to interact with some circuits and not others. The team-of-rivals framework is nicely compatible with neural Darwinism, and emphasizes that Darwinian selection of neural circuitry will tend to strengthen *multiple* circuits—of very different provenance—all of which happen to resonate with a stimulus or task. These circuits are the multiple factions in the brain's congress. For views on the brain as a Darwinian system, see Gerald Edelman, *Neural Darwinism*; Calvin, *How Brains Think*; Dennett, *Consciousness Explained*; or Hayek, *The Sensory Order*.

35 See Weiskrantz, "Outlooks" and *Blindsight*.

36 Technically, reptiles don't see much outside of the immediate reach of their tongues, unless something is moving wildly. So if you're resting on a lounge chair ten feet away from a lizard, you most likely don't exist to him.

37 See, for example, Crick and Koch, "The unconscious homunculus," for use of the term *zombie systems*.

38 A recent finding shows that the Stroop effect can disappear following posthypnotic suggestion. Amir Raz and his colleagues selected a pool of hypnotizable subjects using a completely independent test battery. Under hypnosis, subjects were told that in a later task, they would attend to only ink color. Under these conditions, when the subjects were tested, the Stroop interference essentially vanished. Hypnosis is not a phenomenon that is well understood at the level of the nervous system; nor is it understood why some subjects are more hypnotizable

than others, and what exactly the role of attention, or of reward patterns, might be in explaining the effects. Nevertheless, the data raise intriguing questions about conflict reduction between internal variables, such as a desire to run versus a desire to stay and fight. See Raz, Shapiro, Fan, and Posner, "Hypnotic suggestion."

39 Bem, "Self-perception theory"; Eagleman, "The where and when of intention."

40 Gazzaniga, "The split-brain revisited."

41 Eagleman, Person, and Montague, "A computational role for dopamine." In this paper we constructed a model based on the reward systems in the brain, and ran this model on the same computer game. Astoundingly, the simple model captured the important features of the human strategies, which suggested that people's choices were being driven by surprisingly simple underlying mechanisms.

42 M. Shermer, "Patternicity: Finding meaningful patterns in meaningless noise," *Scientific American*, December 2008.

43 For simplicity, I've related the random-activity hypothesis of dream content, known technically as the activation-synthesis model (Hobson and McCarley, "The brain as a dream state generator"). In fact, there are many theories of dreaming. Freud suggested that dreams are a disguised attempt at wish fulfillment; however, this may be unlikely in the face of, say, the repetitive dreams of post-traumatic stress disorder. Later, in the 1930s, Jung proposed that dreams are compensating for aspects of the personality neglected in waking life. The problem here is that the themes of dreams seem to be the same everywhere, across cultures and generations—themes such as being lost, preparing meals, or being late for an examination—and it's a bit difficult to explain what these things have to do with personality neglect. In general, however, I would like to emphasize that despite the popularity of the activation-synthesis hypothesis in neurobiology circles, there is much about dream content that remains deeply unexplained.

44 Crick and Koch, "Constraints."

45 Tinbergen. "Derived activities."

46 Kelly, *The Psychology of Secrets.*

47 Pennebaker, "Traumatic experience"

48 Petrie, Booth, and Pennebaker, "The immunological effects."

49 To be clear, the team-of-rivals framework, by itself, doesn't solve the whole AI problem. The next difficulty is in learning how to control the subparts, how to dynamically allocate control to expert subsystems, how to arbitrate battles, how to update the system on the basis of recent successes and failures, how to develop a meta-knowledge of how the parts will act when confronted with temptations in the near future, and so on. Our frontal lobes have developed over millions of years using biology's finest tricks, and we still have not teased out the riddles of their circuitry. Nonetheless, understanding the correct architecture from the get-go is our best way forward.

Chapter 6. Why Blameworthiness Is the Wrong Question

1 Lavergne, *A Sniper in the Tower.*

2 Report to Governor, Charles J. Whitman Catastrophe, Medical Aspects, September 8, 1966.

3 S. Brown, and E. Shafer, "An Investigation into the functions of the occipital and temporal lobes of the monkey's brain," *Philosophical Transactions of the Royal Society of London: Biological Sciences* 179 (1888): 303–27.

4 Klüver and Bucy, "Preliminary analysis" This constellation of symptoms, usually accompanied by hypersexuality and hyperorality, is known as Klüver-Bucy syndrome.

5 K. Bucher, R. Myers, and C. Southwick, "Anterior temporal cortex and maternal behaviour in monkey," *Neurology* 20 (1970): 415.

6 Burns and Swerdlow, "Right orbitofrontal tumor."

7 Mendez, et al., "Psychiatric symptoms associated with Alzheimer's disease"; Mendez, et al., "Acquired sociopathy and frontotemporal dementia."

8 M. Leann Dodd, Kevin J. Klos, James H. Bower, Yonas E. Geda, Keith A. Josephs, and J. Eric Ahlskog, "Pathological gambling caused by drugs used to treat Parkinson disease,"*Archives of Neurology* 62, no. 9 (2005): 1377–81.

9 For a solid foundation and clear exposition of the reward systems, see Montague, *Your Brain Is (Almost) Perfect.*

10 Rutter, "Environmentally mediated risks"; Caspi and Moffitt, "Gene–environment interactions."

11 The guilty mind is known as *mens rea*. If you commit the guilty act (*actus reus*) but did not provably have *mens rea*, you are not culpable.

12 Broughton, et al., "Homicidal somnambulism."

13 As of this writing, there have been sixty-eight cases of homicidal somnambulism in North American and European courts, the first one recorded in the 1600s. While we can assume that some fraction of these cases are dishonest pleas, not all of them are. These same considerations of parasomnias have come into courtrooms more recently with sleep scx—for example, rape or infidelity while sleeping—and several cases have been acquitted on these grounds.

14 Libet, Gleason, Wright, and Pearl, "Time"; Haggard and Eimer, "On the relation"; Kornhuber and Deecke, "Changes"; Eagleman, "The where and when of intention"; Eagleman and Holcombe, "Causality"; Soon, et al., "Unconscious determinants of free decisions."

15 Not everyone agrees that Libet's simple test constitutes a meaningful test of free will. As Paul McHugh points out, "What else would one expect when studying a capricious act with neither consequence nor significance to the actor?"

16 Remember, criminal behavior is not entirely about the actor's genes alone. Diabetes and lung disease are influenced by high-sugar foods and elevated air pollution, as well as a genetic predisposition. In the same way, biology and the external environment interact in criminality.

17 Bingham, Preface.

18 See Eagleman and Downar, *Cognitive Neuroscience*.

19 Eadie and Bladin, *A Disease Once Sacred*.

20 Sapolsky, "The frontal cortex."

21 Scarpa and Raine, "The psychophysiology," and Kiehl, "A cognitive neuroscience perspective on psychopathy."

22 Sapolsky, "The frontal cortex."

23 Singer, "Keiner kann anders, als er ist."

24 Note that "abnormal" is meant only in a statistical sense—that is, not the normal way of behaving. The fact that most people behave a certain way is mute on whether that action is correct in a larger moral sense.

It is only a statement about the local laws, customs, and mores of a group of people at a particular time—exactly the same loose constraints by which the "crime" is always defined.

25 See Monahan, "A jurisprudence," or Denno, "Consciousness."

26 A challenge for biological explanations of behavior is that people on the left and right will push their own agendas. See Laland and Brown, *Sense and Nonsense*, as well as O'Hara, "How neuroscience might advance the law." Appropriate caution is of tantamount importance, because biological stories about human behavior have been misused in the past to support agendas. However, past misuse does not mean the biological studies should be abandoned; it only implies that they should be improved.

27 See, for example, Bezdjian, Raine, Baker, and Lynam, "Psychopathic personality," or Raine, *The Psychopathology of Crime*.

28 Note that the lobotomy was considered a successful procedure for noncriminal patients in large part because of the glowing reports of the families. It wasn't immediately appreciated how biased the sources were. Parents would bring in a troubled, loud, dramatic, and troublesome child, and after the surgery the child would be much easier to handle. The mental problems had been replaced by docility. So the feedback was positive. One woman reported of her mother's lobotomy: "She was absolutely violently suicidal beforehand. After the transorbital lobotomy there was nothing. It stopped immediately. It was just peace. I don't know how to explain it to you; it was like turning a coin over. That quick. So whatever [Dr. Freeman] did, he did something right."

As the operation grew in popularity, the age threshold for receiving one went down. The youngest patient to receive the treatment was a twelve-year-old boy named Howard Dully. His stepmother described the behavior that, to her mind, necessitated the operation: "He objects to going to bed but then sleeps well. He does a good deal of daydreaming and when asked about it he says 'I don't know.' He turns the room's lights on when there is broad sunlight outside." And under the ice pick he went.

29 See, for example, Kennedy and Grubin, "Hot-headed or impulsive?", and Stanford and Barratt, "Impulsivity."

30 See LaConte, et al., "Modulating," and Chiu, et al., "Real-time fMRI." Stephen LaConte has been a pioneer in the development of real-time feedback in functional magnetic resonance imaging (fMRI), and he is the mastermind of this work. Pearl Chiu is an expert in psychology and addiction, and she is spearheading the current experiments to use this technology to cure cigarette smokers of their addiction.

31 Imagine a fantasy world in which we could rehabilitate with 100 percent success. Would that mean that systems of punishment would go away? Not entirely. It could be reasonably argued that punishment would still be necessary for two reasons: deterrence of future criminals and the satisfaction of the natural retributive impulse.

32 Eagleman, "Unsolved mysteries."

33 Goodenough, "Responsibility and punishment."

34 Baird and Fugelsang, "The emergence of consequential thought."

35 Eagleman, "The death penalty."

36 Greene and Cohen, "For the law."

37 There are important nuances and subtleties to the arguments presented in this short chapter, and these are developed at more length elsewhere. For those interested in further detail, please see the Initiative on Neuroscience and Law (www.neulaw.org), which brings together neuro-scientists, lawyers, ethicists, and policy makers with the goal of building evidence-based social policy. For further reading, see Eagleman, "Neuroscience and the law," or Eagleman, Correro, and Singh, "Why neuroscience matters."

38 For more about incentive structuring, see Jones, "Law, evolution, and the brain" or Chorvat and McCabe, "The brain and the law."

39 Mitchell and Aamodt, "The incidence of child abuse in serial killers."

40 Eagleman, "Neuroscience and the law."

Chapter 7. Life After the Monarchy

1 Paul, *Annihilation of Man*.

2 Mascall, *The Importance of Being Human*.

3 As for the history of the phrase, the Roman poet Juvenal suggested that "Know thyself" descended straight from heaven (*de caelo*); more sober scholars attribute it to Chilon of Sparta, Heraclitus, Pythagoras,

Socrates, Solon of Athens, Thales of Miletus, or simply popular proverb.

4 Bigelow, "Dr. Harlow's case."

5 *Boston Post*, September 21, 1848, crediting an earlier report from the *Ludlow Free Soil Union* (a Vermont newspaper). The version of the text quoted corrects a confusion in the original report in which the word "diameter" was incorrectly replaced with "circumference." See also Macmillan, *An Odd Kind of Fame*.

6 Harlow, "Recovery."

7 For clarity, I am not compelled by traditional religious stories of the soul. What I mean with the question of a "soul" is something more like a general essence that lives on top of, or outside of, the currently understood biological processes.

8 Pierce and Kumaresan, "The mesolimbic dopamine system."

9 In animal models, researchers will shut down serotonin receptors and demonstrate changes in anxiety and behavior, then restore the receptors and restore normal behavior. For example, see Weisstaub, Zhou and Lira, "Cortical 5-HT2A."

10 Waxman and Geschwind, "Hypergraphia."

11 See Trimble and Freeman, "An investigation," for studies of religiosity in patients with temporal lobe epilepsy, and Devinsky and Lai, "Spirituality," for an overview of epilepsy and religiosity. See d'Orsi and Tinuper, "'I heard voices,'" for the view that the epilepsy of Joan of Arc was a newly described type: idiopathic partial epilepsy with auditory features (IPEAF). See Freemon, "A differential diagnosis," for a historical diagnosis of Muhammad in which he concludes, "Although an unequivocal decision is not possible from existing knowledge, psychomotor or complex partial seizures of temporal lobe epilepsy would be the most tenable diagnosis."

12 I have often wondered whether the promotion of sexual behavior in humans would be the most obvious mechanism for a sexually transmitted virus to advance self-survival. I do not know of any data that support this, but it seems an obvious place to go hunting.

13 There are many more examples of small biological tweaks causing big changes. Patients with herpes simplex encephalitis often get damage

to specific areas of their brains, and show up at the doctor's office with problems in using and understanding the meaning of words—for example, past tense irregulars such as *drive* and *drove*. If you ever intuited that something as impalpable as past tense irregulars is not directly connected to the microscopic knobs, think again. And Creutzfeldt-Jakob disease, a problem caused by abnormally folded proteins called prions, almost always ends in a global dementia characterized by self-neglect, apathy, and irritability. Bizarrely, victims have specific problems with writing, reading, and left-right disorientation. Who would have thought that your sense of left and right had a dependence on the exact folding structure of proteins that are two thousand times smaller than the width of a hair on your head? But there it is.

14 Cummings, "Behavioral and psychiatric symptoms."

15 Sapolsky, "The frontal cortex."

16 See Farah, "Neuroethics."

17 According to one hypothesis of the relationship between schizophrenia and immigration, continual social defeat perturbs dopamine function in the brain. For reviews, see Selten, Cantor-Graae, and Kahn, "Migration," or Weiser, et al., "Elaboration." Thanks to my colleague Jonathan Downar for first bringing this literature to my attention.

18 As of 2008, the U.S. had 2.3 million people behind bars, leading the world in the percentage of its citizens in jail. While society benefits from incarcerating violent repeat offenders, many of those behind bars— such as drug addicts—could be dealt with in a more fruitful manner than imprisonment.

19 Suomi, "Risk, resilience."

20 The genetic change in question lies in the promoter region of the serotonin transporter (5-HTT) gene.

21 Uher and McGuffin, "The moderation," and Robinson, Grozinger, and Whitfield, "Sociogenomics."

22 Caspi, Sugden, Moffitt, et al., "Influence of life stress on depression."

23 Caspi, McClay, Moffitt, et al., "Role of genotype." The genetic change they found was in the promotor region of the gene encoding for monoamine oxidase A (MAOA). MAOA is a molecule that modifies

two neurotransmitter systems critical for mood and emotional regulation: noradrenaline and serotonin.

24 Caspi, Moffitt, Cannon, et al., "Moderation." In this case, the link is a small change in the gene encoding catechol-O-methyltransferase (COMT).

25 Scarpa and Raine, "The psychophysiology of antisocial behaviour."

26 Is it possible that understanding gene–environment interactions could inform preventative approaches? Here's a thought experiment: should we modify the genes once we understand them? We have seen that not everyone who suffers childhood maltreatment follows the path to violence in adulthood. Historically, sociologists have focused on social experiences that might protect some children (for example, can we rescue the child from the abusive home and raise him in a safe and caring environment?). But what has not yet been explored is a protective role of genes—that is, whether genes can protect against environmental insults. While this idea is currently science fiction, it will not be long before someone proposes a gene therapy for such situations: a violence vaccine.

But there's a downside to this sort of intervention: genetic variation is beneficial. We need variation to produce artists, athletes, accountants, architects, and so on. As Stephen Suomi puts it, the "variation seen in certain genes in rhesus monkeys and humans but apparently not in other primate species may actually contribute to their remarkable adaptability and resilience at the species level." In other words, we have a deep ignorance of which genetic combinations end up being most beneficial for a society—and this ignorance provides the firmest argument against genetic intervention. Further, depending on the environment in which one finds oneself, the same set of genes may cause excellence instead of crime. Genes predisposing for aggressiveness may make a talented entrepreneur or CEO; genes predisposing for violence may make a football hero, admired and paid a handsome salary by the population.

27 Kauffman, *Reinventing the Sacred*.

28 Reichenbach, *The Rise of Scientific Philosophy*.

29 One potential sticking point in drawing a relationship between neuro-

science and quantum mechanics is the fact that brain tissue is roughly three hundred degrees Kelvin and is in constant interaction with its immediate environment—these features are not amenable to interesting macroscopic quantum behaviors such as entanglement. Nonetheless, the gap between the two fields is beginning to close, with scientists from both sides making overtures to reach a meaningful hand across the gulf. Moreover, it is now clear that photosynthesis operates with quantum mechanical principles in this same temperature range, which further bespeaks the likelihood that Mother Nature, having figured out how to exploit these tricks in one arena, will exploit them elsewhere. For more on the possibility of quantum effects in the brain, see Koch and Hepp, "Quantum mechanics," or Macgregor, "Quantum mechanics and brain uncertainty."

30 We are sometimes lucky enough to have a hint of what's missing. For example, Albert Einstein felt certain that we were stuck in our psychological filters when it came to understanding the passage of time. Einstein wrote the following to the sister and son of his best friend, Michele Besso, after Besso's death: "Michele has preceded me a little in leaving this strange world. This is not important. For us who are convinced physicists, the distinction between past, present, and future is only an illusion, however persistent." Einstein–Besso correspondence, edited by Pierre Speziali (Paris: Hermann, 1972), 537–39.

Bibliography

Abel, E. 2010. "Influence of names on career choices in medicine." *Names: A Journal of Onomastics*, 58 (2): 65–74.

Ahissar, M., and S. Hochstein. 2004. "The reverse hierarchy theory of visual perceptive learning." *Trends in Cognitive Sciences* 8 (10): 457–64.

Alais, D., and D. Burr. 2004. "The ventriloquist effect results from near-optimal bimodal integration." *Current Biology* 14: 257–62.

Allan, M. D. 1958. "Learning perceptual skills: The Sargeant system of recognition training." *Occupational Psychology* 32: 245–52.

Aquinas, Thomas. 1267–73. *Summa theologiae*, translated by the Fathers of the English Dominican Province. Westminster: Christian Classics, 1981.

Arwas, S., A. Rolnick, and R. E. Lubow. 1989. "Conditioned taste aversion in humans using motion-induced sickness as the US." *Behaviour Research and Therapy* 27 (3): 295–301.

Bach-y-Rita, P. 2004. "Tactile sensory substitution studies." *Annals of the New York Academy of Sciences* 1013: 83–91.

Bach-y-Rita, P., C. C. Collins, F. Saunders, B. White, and L. Scadden. 1969. "Vision substitution by tactile image projection." *Nature* 221: 963–64.

Bach-y-Rita, P., K. A. Kaczmarek, M. E. Tyler, and J. Garcia-Lara. 1998. "Form perception with a 49-point electrotactile stimulus array on the tongue." *Journal of Rehabilitation Research Development* 35: 427–30.

Baird, A. A., and J. A. Fugelsang. 2004. "The emergence of consequential thought: evidence from neuroscience." *Philosophical Transactions of the Royal Society of London B* 359: 1797–1804.

Baker, C. L. Jr., R. F. Hess, and J. Zihl. 1991. "Residual motion perception in a 'motion-blind' patient, assessed with limited-lifetime random dot stimuli." *Journal of Neuroscience* 11 (2): 454–61.

Barkow, J., L. Cosmides, and J. Tooby. 1992. *The Adapted Mind: Evolutionary Psychology and the Generation of Culture.* New York: Oxford University Press.

Bechara, A., A. R. Damasio, H. Damasio, and S. W. Anderson. 1994. "Insensitivity to future consequences following damage to human prefrontal cortex." *Cognition* 50: 7–15.

Bechara, A., H. Damasio, D. Tranel, A. R. Damasio. 1997. "Deciding advantageously before knowing the advantageous strategy." *Science* 275: 1293–95.

Begg, I. M., A. Anas, and S. Farinacci. 1992. "Dissociation of processes in belief: Source recollection, statement familiarity, and the illusion of truth." *Journal of Experimental Psychology* 121: 446–58.

Bell, A. J. 1999. "Levels and loops: The future of artificial intelligence and neuroscience." *Philosophical Transactions of the Royal Society of London B: Biological Sciences* 354 (1392): 2013–20.

Bem, D. J. 1972. "Self-perception theory." In *Advances in Experimental Social Psychology* 6, edited by L. Berkowitz, 1–62. New York: Academic Press.

Benevento, L. A., J. Fallon, B. J. Davis, and M. Rezak. 1977. "Auditory-visual interaction in single cells in the cortex of the superior temporal sulcus and the orbital frontal cortex of the macaque monkey." *Experimental Neurology* 57: 849–72.

Bezdjian, S., A. Raine, L. A. Baker, and D. R. Lynam. 2010. "Psychopathic personality in children: Genetic and environmental contributions." *Psychological Medicine* 20: 1–12.

Biederman, I., and M. M. Shiffrar. 1987. "Sexing day-old chicks." *Journal of Experimental Psychology: Learning, Memory, and Cognition* 13: 640–45.

Bigelow, H. J. 1850. "Dr. Harlow's case of recovery from the passage of an iron bar through the head." *American Journal of the Medical Sciences* 20: 13–22. (Republished in Macmillan, *An Odd Kind of Fame.*)

Bingham, T. 2004. Preface to a special issue on law and brain. *Philosophical Transactions of the Royal Society of London B* 359: 1659.

Blackmore, S. J., G. Brelstaff, K. Nelson, and T. Troscianko. 1995. "Is the richness of our visual world an illusion? Transsaccadic memory for complex scenes." *Perception* 24: 1075–81.

Blakemore, S. J., D. Wolpert, and C. Frith. 2000. "Why can't you tickle yourself?" *Neuroreport* 3 (11): R11–6.

Blake, R., and N. K. Logothetis. 2006. "Visual competition." *Nature Reviews Neuroscience* 3:13–21.

Brandom, R. B. 1998. "Insights and blindspots of reliabilism." *Monist* 81: 371–92.

Brooks, D. N., and A. D. Baddeley. 1976. "What can amnesic patients learn?" *Neuropsychologia* 14: 111-29.

Brooks, R. A. 1986. "A robust layered control system for a mobile robot." *IEEE Journal of Robotics and Automation* April 14–23: RA–2.

Brown, G. 1911. "The intrinsic factors in the act of progression in the mammal." *Proceedings of the Royal Society of London B* 84: 308–19.

Broughton, R., R. Billings, R. Cartwright, D. Doucette, J. Edmeads, M. Edwardh, F. Ervin, B. Orchard, R. Hill, and G. Turrell. 1994. "Homicidal somnambulism: A case study." *Sleep* 17 (3): 253–64.

Bunnell, B. N. 1966. "Amygdaloid lesions and social dominance in the hooded rat." *Psychonomic Science* 6: 93–94.

Burger, J. M., N. Messian, S. Patel, A. del Prado, and C. Anderson. 2004. "What a coincidence! The effects of incidental similarity on compliance." *Personality and Social Psychology Bulletin* 30: 35–43.

Burns, J. M., and R. H. Swerdlow. 2003. "Right orbitofrontal tumor with pedophilia symptom and constructional apraxia sign." *Archives of Neurology* 60 (3): 437–40.

Calvert, G. A., E. T. Bullmore, M. J. Brammer, et al. 1997. "Activation of auditory cortex during silent lipreading." *Science* 276: 593–96.

Calvin, W. H. 1996. *How Brains Think: Evolving Intelligence, Then and Now*. New York: Basic Books.

Carter, R. 1998. *Mapping the Mind*. Berkeley: University of California Press.

Caspi, A., J. McClay, and T. E. Moffitt, et al. 2002. "Role of genotype in the cycle of violence in maltreated children." *Science* 297: 851.

Caspi, A., K. Sugden, T. E. Moffitt, et al. 2003. "Influence of life stress on depression: Moderation by a polymorphism in the 5-HTT gene." *Science* 301: 386.

Caspi, A., T. E. Moffitt, M. Cannon, et al. 2005. "Moderation of the effect of adolescent-onset cannabis use on adult psychosis by a functional polymorphism in the COMT gene: Longitudinal evidence of a gene environment interaction." *Biological Psychiatry* 57: 1117–27.

Caspi, A., and T. E. Moffitt. 2006. "Gene–environment interactions in psychiatry: Joining forces with neuroscience." *Nature Reviews Neuroscience* 7: 583–90.

Cattell, J. M. 1886. "The time taken up by cerebral operations." *Mind* 11: 220–42.

Cattell, J. M. 1888. "The psychological laboratory at Leipsic." *Mind* 13: 37–51.

Chance, B. 1962. *Ophthalmology*. New York: Hafner.

Chiu, P., B. King Casas, P. Cinciripini, F. Versace, D. M. Eagleman, J. Lisinski, L. Lindsey, and S. LaConte. 2009. "Real-time fMRI modulation of craving and control brain states in chronic smokers." Abstract presented at the Society for Neuroscience, Chicago, IL.

Chorvat, T., and K. McCabe. 2004. "The brain and the law." *Philosophical Transactions of the Royal Society of London B* 359: 1727–36.

Cleeremans, A. 1993. *Mechanisms of Implicit Learning*. Cambridge, MA: MIT Press.

Clifford, C. W., and M. R. Ibbotson. 2002. "Fundamental mechanisms of visual motion detection: Models, cells and functions." *Progress in Neurobiology* 68 (6): 409–37.

Cohen, J. D. 2005. "The vulcanization of the human brain: A neural perspective on interactions between cognition and emotion." *Journal of Economic Perspectives* 19 (4): 3–24.

Cohen, N. J., H. Eichenbaum, B. S. Deacedo, and S. Corkin. 1985. "Different memory systems underlying acquisition of procedural and declarative knowledge." *Annals of the New York Academy of Sciences* 444: 54–71.

Collett, T. S., and M. F. Land. 1975. "Visual control of flight behaviour in the hoverfly *Syritta pipiens*." *Journal of Comparative Physiology* 99: 1–66.

Cosmides, L., and J. Tooby. 1992. *Cognitive Adaptions for Social Exchange*. New York: Oxford University Press.

Crick, F. H. C., and C. Koch. 1998. "Constraints on cortical and thalamic projections: The no-strong-loops hypothesis." *Nature* 391 (6664): 245–50.

———. 2000. "The unconscious homunculus. " In *The Neuronal Correlates of Consciousness*, edited by T. Metzinger, 103–110. Cambridge, MA: MIT Press.

Cui, X., D. Yang, C. Jeter, P. R. Montague, and D. M. Eagleman. 2007. "Vividness of mental imagery: Individual variation can be measured objectively." *Vision Research* 47: 474–78.

Cummings, J. 1995. "Behavioral and psychiatric symptoms associated with Huntington's disease." *Advances in Neurology* 65: 179–88.

Cytowic, R. E. 1998. *The Man Who Tasted Shapes*. Cambridge, MA: MIT Press.

Cytowic, R. E., and D. M. Eagleman. 2009. *Wednesday Is Indigo Blue: Discovering the Brain of Synesthesia*. Cambridge, MA: MIT Press.

Damasio, A. R. 1985. "The frontal lobes." In *Clinical Neuropsychology*, edited by K. M. Heilman and E. Valenstein, 339–75. New York: Oxford University Press.

———. 1994. *Descartes' Error: Emotion, Reason and the Human Brain*. New York: Putnam.

———. 1999. *The Feeling of What Happens: Body and Emotion in the Making of Consciousness*. New York: Houghton Mifflin Harcourt.

Damasio, A. R., B. J. Everitt, and D. Bishop. 1996. "The somatic marker hypothesis and the possible functions of the prefrontal cortex." *Philosophical Transactions: Biological Sciences* 351 (1346): 1413–20.

D'Angelo, F. J. 1986. "Subliminal seduction: An essay on the rhetoric of the unconscious." *Rhetoric Review* 4 (2): 160–71.

de Gelder, B., K. B. Bocker, J. Tuomainen, M. Hensen, and J. Vroomen. 1999. "The combined perception of emotion from voice and face: Early interaction revealed by human electric brain responses." *Neuroscience Letters* 260: 133–36.

Dennett, D. C. 1991. *Consciousness Explained*. Boston: Little, Brown and Company.

Dennett, D. C. 2003. *Freedom Evolves*. New York: Viking Books.

Denno, D. W. 2009. "Consciousness and culpability in American criminal law." *Waseda Proceedings of Comparative Law*, vol. 12, 115–26.

Devinsky, O., and G. Lai. 2008. "Spirituality and religion in epilepsy." *Epilepsy Behav*iour 12 (4): 636–43.

Diamond, J. 1999. *Guns, Germs, and Steel*. New York: Norton.

d'Orsi, G. and P. Tinuper. 2006. "'I heard voices . . .': From semiology, a historical review, and a new hypothesis on the presumed epilepsy of Joan of Arc." *Epilepsy and Behaviour* 9 (1): 152–57.

Dully, H. and C. Fleming. 2007. *My Lobotomy*. New York: Crown.

Eadie, M. and P. Bladin. 2001. *A Disease Once Sacred: A History of the Medical Understanding of Epilepsy*. New York: Butterworth-Heinemann.

Eagleman, D. M. 2001. "Visual illusions and neurobiology." *Nature Reviews Neuroscience* 2 (12): 920–26.

———. 2004. "The where and when of intention." *Science* 303: 1144–46.

———. 2005. "The death penalty for adolescents." Univision television interview. *Too Young To Die?* May 24.

———. 2005. "Distortions of time during rapid eye movements." *Nature Neuroscience* 8 (7): 850–51.

———. 2006. "Will the internet save us from epidemics?" *Nature* 441 (7093): 574.

———. 2007. "Unsolved mysteries of the brain." *Discover* August.

———. 2008. "Human time perception and its illusions." *Current Opinion in Neurobiology* 18 (2): 131–36.

———. 2008. "Neuroscience and the law." *Houston Lawyer* 16 (6): 36–40.

———. 2008. "Prediction and postdiction: Two frameworks with the goal of delay compensation." *Brain and Behavioral Sciences* 31 (2): 205–06.

———. 2009. "America on deadline." *New York Times*. December 3.

———. 2009. "Brain time." In *What's Next: Dispatches from the Future of Science*, edited by M. Brockman. New York: Vintage Books. (Reprinted at Edge.org.)

———. 2009. "The objectification of overlearned sequences: A large-scale analysis of spatial sequence synesthesia." *Cortex* 45 (10): 1266–77.

———. 2009. "Silicon immortality: Downloading consciousness into computers." In *What Will Change Everything?* edited by J. Brockman. New York: Vintage Books. (Originally printed at Edge.org.)

———. 2009. *Sum: Tales from the Afterlives*. Edinburgh: Canongate Books.

———. 2009. "Temporality, empirical approaches." In *The Oxford Companion to Consciousness*. Oxford, UK: Oxford University Press.

———. 2010. "Duration illusions and predictability." In *Attention and Time*, edited by J. T. Coull and K. Nobre. New York: Oxford University Press.

———. 2010. "How does the timing of neural signals map onto the timing of perception?" In *Problems of Space and Time in Perception and Action*, edited by R. Nijhawan. Cambridge, UK: Cambridge University Press.

———. 2010. "Synaesthesia." *British Medical Journal* 340: b4616.

———. 2011. "The Brain on Trial." *The Atlantic*.

———. 2012. *Live-Wired: The Shape Shifting Brain*. Oxford: Oxford University Press.

Eagleman, D. M., M. A. Correro, and J. Singh. 2010. "Why neuroscience matters for a rational drug policy." *Minnesota Journal of Law, Science and Technology*. In press.

Eagleman, D. M., and J. Downar. 2011. *Cognitive Neuroscience: A Principles-Based Approach*. New York: Oxford University Press, forthcoming.

Eagleman, D. M., and M. A. Goodale. 2009. "Why color synesthesia involves more than color." *Trends in Cognitive Sciences* 13 (7): 288–92.

Eagleman, D. M. and A. O. Holcombe. 2002. "Causality and the perception of time." *Trends in Cognitive Sciences*. 6 (8): 323–25.

Eagleman, D. M., J. E. Jacobson, and T. J. Sejnowski. 2004. "Perceived luminance depends on temporal context." *Nature* 428 (6985): 854–56.

Eagleman, D. M., A. D. Kagan, S. N. Nelson, D. Sagaram, and A. K. Sarma 2007. "A standardized test battery for the study of synesthesia." *Journal of Neuroscience Methods* 159: 139–45.

Eagleman, D. M., and P. R. Montague. 2002. "Models of learning and memory." In *Encyclopedia of Cognitive Science*. London: Macmillan Press.

Eagleman, D. M., and V. Pariyadath. 2009. "Is subjective duration a signature of coding efficiency?" *Philosophical Transactions of the Royal Society* 364 (1525): 1841–51.

Eagleman, D. M., C. Person, and P. R. Montague. 1998. "A computational role for dopamine delivery in human decision-making." *Journal of Cognitive Neuroscience* 10 (5): 623–30.

Eagleman, D. M., and T. J. Sejnowski. 2000. "Motion integration and postdiction in visual awareness." *Science* 287 (5460): 2036–38.

———. 2007. "Motion signals bias position judgments: A unified explanation for the flash-lag, flash-drag, flash-jump and Frohlich effects." *Journal of Vision* 7 (4): 1–12.

Eagleman, D. M., P. U. Tse, P. Janssen, A. C. Nobre, D. Buonomano, and A. O. Holcombe. 2005. "Time and the brain: How subjective time relates to neural time." *Journal of Neuroscience.* 25 (45): 10369–71.

Ebbinghaus, H. (1885) 1913. *Memory: A Contribution to Experimental Psychology*, translated by Henry A. Ruger & Clara E. Bussenius. New York: Teachers College, Columbia University.

Edelman, G. M. 1987. *Neural Darwinism. The Theory of Neuronal Group Selection.* New York: Basic Books.

Edelman, S. 1999. *Representation and Recognition in Vision.* Cambridge, MA: MIT Press.

———. 2008. *Computing the Mind: How the Mind Really Works.* Oxford: Oxford University Press.

Elliott, R., R. J. Dolan, and C. D. Frith. 2000. "Dissociable functions in the medial and lateral orbitofrontal cortex: Evidence from human neuroimaging studies." *Cerebral Cortex* 10 (3): 308–17.

Emerson, R. W. (1883) 1984. *Emerson in His Journals.* Reprint, Cambridge, MA: Belknap Press of Harvard University Press.

Ernst, M. O., and M. S. Banks. 2002. "Humans Integrate visual and haptic information in a statistically optimal fashion." *Nature* 415: 429–33.

Evans, J. S. 2008. "Dual-processing accounts of reasoning, judgment, and social cognition." *Annual Review of Psychology* 59: 255–78.

Exner, S. 1875. "Experimentelle Untersuchung der einfachsten psychischen Processe." *Pflüger's Archive: European Journal of Physiology* 11: 403–32.

Farah, M. J. 2005. "Neuroethics: The practical and the philosophical." *Trends in Cognitive Sciences* 9: 34–40.

Faw, B. 2003. "Pre-frontal executive committee for perception, working memory, attention, long-term memory, motor control, and thinking: A tutorial review." *Consciousness and Cognition* 12 (1): 83–139.

Festinger, L. 1964. *Conflict, Decision, and Dissonance.* Palo Alto, CA: Stanford University Press.

Fisher, H. 1994. *Anatomy of Love: The Natural History of Mating, Marriage and Why We Stray.* New York: Random House.

Frederick, S., G. Loewenstein, and T. O'Donoghue. 2002. "Time discounting and time preference: A critical review." *Journal of Economic Lite*rature 40: 351.

Freeman, J. B., N. Ambady, N. O. Rule, and K. L. Johnson. 2008. "Will a category cue attract you? Motor output reveals dynamic competition across person construal." *Journal of Experimental Psychology: General* 137 (4): 673–90.

Freemon, F. R. 1976. "A differential diagnosis of the inspirational spells of Muhammad the prophet of Islam." *Epilepsia.* 17 (4): 423–27.

Freud, S. 1927. *The Standard Edition of the Complete Psychological Works of Sigmund Freud.* Volume 21, *The Future of an Illusion.* Translated by James Strachey. London: Hogarth Press, 1968.

Freud, S. and J. Breuer. 1895. *Studien über Hysterie (Studies on Hysteria).* Leipzig: Franz Deuticke.

Friedman, R. S., D. M. McCarthy, J. Forster, and M. Denzler. 2005. "Automatic effects of alcohol cues on sexual attraction." *Addiction* 100 (5): 672–81.

Frith, C. and R. J. Dolan. 1997. "Brain mechanisms associated with top-down processes in perception." *Philosophical Transactions of the Royal Society of London B: Biological Sciences* 352 (1358): 1221–30.

Fuller, J. L., H. E. Rosvold, and K. H. Pribram. 1957. "The effect on affective and cognitive behavior in the dog of lesions of the pyriformamygdala-hippocampal complex." *Journal of Comparative and Physiological Psychology* 50 (1): 89–96.

Fusi, S., P. J. Drew, and L. F. Abbott. 2005. "Cascade models of synaptically stored memories." *Neuron* 45 (4): 599–611.

Garland, B., ed. 2004. *Neuroscience and the Law: Brain, Mind, and the Scales of Justice.* New York: Dana Press.

Gazzaniga, M. S. 1998. "The split-brain revisited." *Scientific American* 279 (1): 35–9.

Gebhard, J. W. and G. H. Mowbray. 1959. "On discriminating the rate of visual flicker and auditory flutter." *American Journal of Experimental Psychology* 72: 521–28.

Gloor, P. 1960. *Amygdala*. In *J. Field Handbook of Physiology*, edited by H. W. Magoun and V. E. Hall, vol. 2, 1395–1420. Washington: American Physiological Society.

Goldberg, E. 2001. *The Executive Brain: Frontal Lobes and the Civilized Mind*. New York: Oxford University Press.

Goodenough, O. R. 2004. "Responsibility and punishment: Whose mind? A response." *Philosophical Transactions of the Royal Society of London B* 359: 1805–09.

Goodwin, D. Kearns. 2005. *Team of Rivals: The Political Genius of Abraham Lincoln*. New York: Simon & Schuster.

Gould, S. J. 1994. "The evolution of life on Earth." *Scientific American* 271 (4): 91.

Graf, P. and D. L. Schacter. 1985. "Implicit and explicit memory for new associations in normal and amnesic subjects." *Journal of Experimental Psychology: Learning, Memory, and Cognition* 11: 501–518.

———. 1987. "Selective effects of interference on implicit and explicit memory for new associations." *Journal of Experimental Psychology: Learning, Memory, and Cognition* 13: 45–53.

Greene, J. and J. Cohen. 2004. "For the law, neuroscience changes nothing and everything." *Philosophical Transactions of the Royal Society of London B* 359: 1775–85.

Greene, J., L. Nystrom, A. Engell, J. Darley, and J. Cohen. 2004. "The neural bases of cognitive conflict and control in moral judgment." *Neuron* 44 (2): 389–400.

Greenwald, A. G., D. E. McGhee, and J. K. L. Schwartz. 1998. "Measuring individual differences in implicit cognition: The implicit association test." *Journal of Personality and Social Psychology* 74: 1464–80.

Grossberg, S. 1980. "How does a brain build a cognitive code?" *Psychological Review* 87 (1):1–51.

Grush, R. "The emulation theory of representation: Motor control, imagery, and perception." *Behavioral and Brain Sciences* 27: 377–442.

Gutnisky, D. A., B. J. Hansen, B. F. Iliescu, and V. Dragoi. 2009. "Attention alters visual plasticity during exposure-based learning." *Current Biology* 19 (7): 555–60.

Haggard, P. and M. Eimer. 1999. "On the relation between brain potentials and the awareness of voluntary movements." *Experimental Brain Research* 126: 128–33.

Haidt, J. 2001. "The emotional dog and its rational tail: A social intuitionist approach to moral judgment." *Psychological Review* 108: 814–34.

———.2007. "The new synthesis in moral psychology." *Science* 316 (5827): 998.

Harlow, J. M. 1868. "Recovery from the passage of an iron bar through the head." *Publications of the Massachusetts Medical Society* 2: 327–47. (Republished in Macmillan, *An Odd Kind of Fame*.)

Harnad, S. 1996. "Experimental analysis of naming behavior cannot explain naming capacity." *Journal of the Experimental Analysis of Behavior* 65: 262–64.

Hasher, L., D. Goldstein, and T. Toppino. 1977. "Frequency and the conference of referential validity." *Journal of Verbal Learning and Verbal Behavior* 16: 107–12.

Hassin, R., J. S. Uleman, and J. A. Bargh. 2004. *The New Unconscious*. New York: Oxford University Press.

Hawkins, J., with S. Blakeslee. 2005. *On Intelligence*. New York: Henry Holt.

Hayek, F. A. 1952. *The Sensory Order: An Inquiry into the Foundations of Theoretical Psychology*. London: Routledge & Kegan Paul.

Heidelberger, M. 2004. *Nature from Within: Gustav Theodor Fechner and His Psychophysical Worldview*. Translated by C. Klohr. Pittsburgh, PA: University of Pittsburgh Press.

Helmholtz, H. von. 1857–67. *Handbuch der physiologischen Optik*. Leipzig: Voss.

Herbart, J. F. 1961. *Psychology as a Science, Newly Founded On Experience, Metaphysics and Mathematics*. In *Classics in Psychology*, edited by Thorne Shipley. New York: Philosophical Library.

Hobson, J. A. and R. McCarley. 1977. "The brain as a dream state generator: An activation-synthesis hypothesis of the dream process." *American Journal of Psychiatry* 134: 1335–48.

Holcombe, A. O., N. Kanwisher, and A. Treisman. 2001. "The midstream order deficit." *Perception and Psychophysics* 63 (2): 322–29.

Honderich, T. 2002. *How Free Are You? The Determinism Problem.* New York: Oxford University Press.

Horsey, R. 2002. *The Art of Chicken Sexing.* University College London Working Papers in Linguistics.

Huxley, J. 1946. *Rationalist Annual,* 87. London: C. A. Watts.

Ingle, D. 1973. "Two visual systems in the frog." *Science* 181: 1053–55.

Jacobs, R., M. I. Jordan, S. J. Nowlan, and G. E. Hinton. 1991. "Adaptive mixtures of local experts." *Neural Computation* 3: 79–87.

Jacoby, L. L., and D. Witherspoon. 1982. "Remembering without awareness." *Canadian Journal of Psychology* 32: 300–24.

James, W. 1890. *Principles of Psychology.* New York: Henry Holt.

Jameson, K. A. 2009. "Tetrachromatic color vision." In *The Oxford Companion to Consciousness,* edited by P. Wilken, T. Bayne and A. Cleeremans. Oxford: Oxford University Press.

Jaynes, J. 1976. *The Origin of Consciousness in the Breakdown of the Bicameral Mind.* Boston: Houghton Mifflin.

Johnson, M. H., and J. Morton. 1991. "CONSPEC and CONLERN: A two-process theory of infant face recognition." *Psychological Review* 98 (2): 164–81.

Jones, J. T., B. W. Pelham, M. Carvallo, and M. C. Mirenberg. 2004. "How do I love thee? Let me count the Js: Implicit egotism and interpersonal attraction." *Journal of Personality and Social Psychology* 87 (5): 665–83.

Jones, O. D. 2004. "Law, evolution, and the brain: Applications and open questions." *Philosophical Transactions of the Royal Society of London Series B: Biological Sciences* 359: 1697–1707.

Jordan, M. I., and R. A. Jacobs. 1994. "Hierarchical mixtures of experts and the EM algorithm." *Neural Computation* 6: 181–214.

Jung, C. G., and A. Jaffé. 1965. *Memories, Dreams, Reflections.* New York: Random House.

Kahneman, D., and S. Frederick. 2002. "Representativeness revisited: Attribute substitution in intuitive judgment." In *Heuristics and Biases,* edited by T. Gilovich, D. Griffin, and D. Kahneman, 49–81. New York: Cambridge University Press.

Kauffman, S. A. 2008. *Reinventing the Sacred: A New View of Science, Reason, and Religion*. New York: Basic Books.

Kawato, M. 1999. "Internal models for motor control and trajectory planning." *Current Opinion in Neurobiology* 9: 718–27.

Kawato, M., K. Furukawa, and R. Suzuki. 1987. "A hierarchical neural-network model for control and learning of voluntary movement." *Biological Cybernetics* 57: 169–185.

Kelly, A. E. 2002. *The Psychology of Secrets*. The Plenum Series in Social/Clinical Psychology. New York: Plenum.

Kennedy, H. G., and D. H. Grubin. 1990. "Hot-headed or impulsive?" *British Journal of Addiction* 85 (5): 639–43.

Kersten, D., D. C. Knill, P. Mamassian, and I. Bülthoff. 1996. "Illusory motion from shadows." *Nature* 279 (6560): 31.

Key, W. B. 1981. *Subliminal seduction: Ad Media's Manipulation of a Not So Innocent America*. New York: New American Library.

Kidd, B. 1894. *Social Evolution*. New York and London: Macmillan.

Kiehl, K. A. 2006. "A cognitive neuroscience perspective on psychopathy: Evidence for paralimbic system dysfunction." *Psychiatry Research* 142 (2–3): 107–28.

Kitagawa, N., and S. Ichihara. 2002. "Hearing visual motion in depth." *Nature* 416: 172–74.

Kling, A., and L. Brothers. 1992. "The amygdala and social behavior." In *Neurobiological Aspects of Emotion, Memory, and Mental Dysfunction*, edited by J. Aggleton. New York: Wiley-Liss.

Klüver, H., and P. C. Bucy. 1939. "Preliminary analysis of functions of the temporal lobes in monkeys." *Archives of Neurology and Psychiatry* 42: 979–1000.

Koch, C., and K. Hepp. 2006. "Quantum mechanics in the brain." *Nature* 440 (7084): 611.

Kornhuber, H. H., and L. Deecke. 1965. "Changes in brain potentials with willful and passive movements in humans: The readiness potential and reafferent potentials." *Pflüger's Archive* 284: 1–17.

Kosik, K. S. 2006. "Neuroscience gears up for duel on the issue of brain versus deity." *Nature* 439 (7073): 138.

Kurson, R. 2007. *Crashing Through*. New York: Random House.

LaConte, S., B. King Casas, J. Lisinski, L. Lindsey, D. M. Eagleman, P. M. Cinciripini, F. Versace, and P. H. Chiu. 2009. "Modulating real time fMRI neurofeedback interfaces via craving and control in chronic smokers." Abstract presented at the Organization for Human Brain Mapping, San Francisco, CA.

Lacquaniti, F., M. Carrozzo, and N. A. Borghese. 1993. "Planning and control of limb impedance." In *Multisensory Control of Movement*, edited by A. Berthoz. Oxford: Oxford University Press.

Laland, K. L., and G. R. Brown. 2002. *Sense and Nonsense: Evolutionary Perspectives on Human Behavior.* New York: Oxford University Press.

Lanchester, B. S., and R. F. Mark. 1975. "Pursuit and prediction in the tracking of moving food by a teleost fish (*Acanthaluteres spilomelanurus*). *Journal of Experimental Biology* 63 (3): 627–45.

Lavergne, G. M. 1997. *A Sniper in the Tower: The True Story of the Texas Tower Massacre.* New York: Bantam.

Leibniz, G. W. 1679. *De Progressione Dyadica, Pars I.* (Manuscript dated 15 March 1679, published in facsimile (with German translation) in *Herrn von Leibniz' Rechnung mit Null und Einz*, edited by Erich Hochstetter and Hermann-Josef Greve, pp. 46–47. Berlin: Siemens Aktiengesellschaft, 1966. English translation by Verena Huber-Dyson, 1995.

Leibniz, G. W. 1704, published 1765. *Nouveaux essais sur l'entendement humain.* Published in English in 1997 as *New Essays on Human Understanding*, translated by Peter Remnant and Jonathan Bennett. Cambridge, UK: Cambridge University Press.

Levin, D. T., and D. J. Simons. 1997. "Failure to detect changes to attended objects in motion pictures." *Psychonomic Bulletin & Review* 4 (4): 501–506.

Lewis, J. W., M. S. Beauchamp, and E. A. DeYoe. 2000. "A comparison of visual and auditory motion processing in human cerebral cortex." *Cerebral Cortex* 10 (9): 873–88.

Liberles, S. D., and L. B. Buck. 2006. "A second class of chemosensory receptors in the olfactory epithelium." *Nature* 442, 645–50.

Libet, B., Gleason, C. A., Wright, E. W., and Pearl, D. K. 1983. "Time of conscious intention to act in relation to onset of cerebral activity (readiness-potential): The unconscious initiation of a freely voluntary act." *Brain* 106: 623–42.

Libet, B. 2000. *The Volitional Brain: Towards a Neuroscience of Free Will.* Charlottesville, VA: Imprint Academic.

Lim, M., Z. Wang, D. Olazabal, X. Ren, E. Terwilliger, and L. Young. 2004. "Enhanced partner preference in a promiscuous species by manipulating the expression of a single gene." *Nature* 429: 754–57.

Livnat, A., and N. Pippenger. 2006. "An optimal brain can be composed of conflicting agents." *Proceedings of the National Academy of Sciences* 103: 3198–3202

Llinas, R. 2002. *I of the Vortex.* Boston: MIT Press.

Loe, P. R., and L. A. Benevento. 1969. "Auditory-visual interaction in single units in the orbito-insular cortex of the cat." *Electroencephalography and Clinical Neurophysiology* 26: 395–98.

Macaluso, E., C. D. Frith, and J. Driver. 2000. "Modulation of human visual cortex by crossmodal spatial attention." *Science* 289: 1206–8.

Macgregor, R. J. 2006. "Quantum mechanics and brain uncertainty." *Journal of Integrative Neuroscience* 5 (3): 373–80.

Macknik, S. L., M. King, J. Randi, et al. 2008. "Attention and awareness in stage magic: Turning tricks into research." *Nature Reviews Neuroscience* 9: 871–879.

MacKay, D. M. 1956. "The epistemological problem for automata." In *Automata Studies*, edited by C. E. Shannon and J. McCarthy, 235–51. Princeton: Princeton University Press.

MacKay, D. M. 1957. "Towards an information-flow model of human behavior." *British Journal of Psychology* 47: 30–43.

MacLeod, D. I. A., and I. Fine. 2001. "Vision after early blindness." Abstract. *Journal of Vision* 1 (3): 470, 470a.

Macmillan, M. 2000. *An Odd Kind of Fame: Stories of Phineas Gage.* Cambridge, MA: MIT Press.

Macuga, K. L., A. C. Beall, J. W. Kelly, R. S. Smith, J. M. Loomis. 2007. "Changing lanes: Inertial cues and explicit path information facilitate steering performance when visual feedback is removed." *Experimental Brain Research* 178 (2): 141–50.

Manning, J. T., D. Scutt, G. H. Whitehouse, S. J. Leinster, J. M. Walton. 1996. "Asymmetry and the menstrual cycle in women." *Ethology and Sociobiology* 17, 129–43.

Marlowe, W. B., E. L. Mancall, and J. J. Thomas. 1975. "Complete Klüver-Bucy syndrome in man." *Cortex* 11 (1): 53–59.

Marr, D. 1982. *Vision*. San Francisco: W. H. Freeman.

Mascall, E. L. 1958. *The Importance of Being Human*. New York: Columbia University.

Massaro, D. W. 1985. "Attention and perception: An information-integration perspective. *Acta Psychologica (Amsterdam)* 60: 211–43.

Mather, G., A. Pavan, G. Campana, and C. Casco. 2008. "The motion aftereffect reloaded." *Trends in Cognitive Sciences* 12 (12): 481–87.

Mather, G., F. Verstraten, and S. Anstis. 1998. *The Motion Aftereffect: A Modern Perspective*. Cambridge, MA: MIT Press.

McBeath, M. K., D. M. Shaffer, and K. M. Kaiser. 1995. "How baseball out-fielders determine where to run to catch fly balls." *Science* 268: 569–73.

McClure, S. M., D. I. Laibson, G. Loewenstein, and J. D. Cohen. 2004. "Separate neural systems value immediate and delayed monetary rewards." *Science* 306 (5695): 503–07.

McClure, S. M., M. M. Botvinick, N. Yeung, J. D. Greene, and J. D. Cohen. 2007. "Conflict monitoring in cognition-emotion competition." In *Handbook of Emotion Regulation*, edited by J. J. Gross. New York: The Guilford Press.

McGurk, H., and J. MacDonald. 1976. "Hearing lips and seeing voices." *Nature* 264: 746–48.

McIntyre, J., M. Zago, A. Berthoz, and F. Lacquaniti. 2001. "Does the brain model Newton's laws?" *Nature Neuroscience* 4: 693–94.

Mehta, B., and S. Schaal. 2002. "Forward models in visuomotor control." *Journal of Neurophysiology* 88: 942–53.

Meltzoff, A. N. 1995. "Understanding the intentions of others: Re-enactment of intended acts by 18-month-old children." *Developmental Psychology* 31: 838–50.

Mendez, M. F., R. J. Martin, K. A. Amyth, P. J. Whitehouse. 1990. "Psychiatric symptoms associated with Alzheimer's disease." *Journal of Neuropsychiatry* 2: 28–33.

Mendez, M. F., A. K. Chen, J. S. Shapira, and B. L. Miller. 2005. "Acquired sociopathy and frontotemporal dementia." *Dementia and Geriatric Cognitive Disorders* 20 (2–3): 99–104.

Meredith, M. A., J. W. Nemitz, and B. E. Stein. 1987. "Determinants of multisensory integration in superior colliculus neurons. I. Temporal factors." *Journal of Neuroscience* 7: 3215–29.

Mesulam, M. 2000. *Principles of Behavioral and Cognitive Neurology.* New York: Oxford University Press.

Miall, R. C., and D. M. Wolpert. 1996. "Forward models for physiological motor control." *Neural Network* 9 (8): 1265–79.

Miller, N. E. 1944. "Experimental studies in conflict." In *Personality and the Behavior Disorders*, edited by J. Hunt, vol. 1, 431–65.

Milner, D., and M. Goodale. 1995. *The Visual Brain in Action.* Oxford: Oxford University Press.

Minsky, M. 1986. *Society of Mind.* New York: Simon and Schuster.

Mitchell, H., and M. G. Aamodt. 2005. "The incidence of child abuse in serial killers." *Journal of Police and Criminal Psychology* 20 (1): 40–47.

Mocan, N. H., and R. K. Gittings. 2008. "The impact of incentives on human behavior: Can we make it disappear? The case of the death penalty." Working paper, National Bureau of Economic Research.

Moffitt, T. E., and B. Henry. 1991. "Neuropsychological studies of juvenile delinquency and juvenile violence." In *Neuropsychology of Aggression*, edited by J. S. Milner. Boston: Kluwer.

Moles, A., B. L. Kieffer and F. R. D'Amato. 2004. "Deficit in attachment behavior in mice lacking the mu-opioid receptor gene." *Science* 304 (5679): 1983–86.

Monahan, J. 2006. "A jurisprudence of risk assessment: Forecasting harm among prisoners, predators, and patients." *Virginia Law Review* 92 (33): 391–417.

Montague, P. R. 2008. *Your Brain Is (Almost) Perfect: How We Make Decisions.* New York: Plume.

Montague, P. R., P. Dayan, C. Person, and T. J. Sejnowski. 1995. "Bee foraging in uncertain environments using predictive Hebbian learning." *Nature* 377: 725–28

Morse, S. 2004. "New neuroscience, old problems." In *Neuroscience and the Law: Brain, Mind, and the Scales of Justice*, edited by B. Garland. New York: Dana Press.

Mumford, D. 1992. "On the computational architecture of the neocortex. II. The role of cortico-cortical loops." *Biological Cybernetics* 66 (3): 241–51.

Nagel, T. 1986. *The View from Nowhere*. New York: Oxford University Press.

Nakayama, K., and C. W. Tyler. 1981. "Psychophysical isolation of movement sensitivity by removal of familiar position cues." *Vision Research* 21 (4): 427–33.

Niedenthal, P. M. 2007. "Embodying emotion." *Science* 316 (5827): 1002.

Noë, A. 2005. *Action in Perception*. Cambridge, MA: MIT Press.

Norretranders, T. 1992. *The User Illusion: Cutting Consciousness Down to Size*. New York: Penguin Books.

Novich S, Cheng S, Eagleman D. M. 2011. "Is synaesthesia one condition or many? A large-scale analysis reveals subgroups." *Journal of Neuropsychology*. 5(2):353-71.

O'Hara, E. A., and D. Yarn. 2002. "On apology and consilience." *Washington Law Review* 77: 1121.

O'Hara, E. A. 2004. "How neuroscience might advance the law." *Philosophical Transactions of the Royal Society B* 359: 1677–84.

O'Hare, D. 1999. "Introduction to human performance in general aviation." In *Human performance in general aviation*, edited by D. O'Hare, 3–10. Aldershot, UK: Ashgate.

O'Regan, J. K. 1992. "Solving the real mysteries of visual perception: The world as an outside memory." *Canadian Journal of Psychology* 46: 461–88.

Pariyadath, V., and D. M. Eagleman. 2007. "The effect of predictability on subjective duration." *PLoS One* 2 (11): e1264.

Paul, L. 1945. *Annihilation of Man*. New York: Harcourt Brace.

Pearson, H. 2006. "Mouse data hint at human pheromones: Receptors in the nose pick up subliminal scents." *Nature* 442: 95.

Pelham, B. W., M. Carvallo, and J. T. Jones. 2005. "Implicit egotism." *Current Directions in Psychological Science* 14: 106–10.

Pelham, B. W., S. L. Koole, C. D. Hardin, J. J. Hetts, E. Seah, and T. DeHart, 2005. "Gender moderates the relation between implicit and explicit self-esteem." *Journal of Experimental Social Psychology.* 41: 84–89.

Pelham, B. W., M. C. Mirenberg, and J.T. Jones. 2002. "Why Susie sells seashells by the seashore: Implicit egotism and major life decisions." *Journal of Personality and Social Psychology* 82: 469–87.

Pennebaker, J. W. 1985. "Traumatic experience and psychosomatic disease: Exploring the roles of behavioral inhibition, obsession, and confiding." *Canadian Psychology* 26: 82–95.

Penton-Voak, I. S., D. I. Perrett, D. Castles, M. Burt, T. Koyabashi, and L. K. Murray. 1999. "Female preference for male faces changes cyclically." *Nature* 399: 741–42.

Petrie, K. P., R. J. Booth, and J. W. Pennebaker. 1998. "The immunological effects of thought suppression." *Journal of Personality and Social Psychology* 75: 1264–72.

Pierce, R. C., and V. Kumaresan. 2006. "The mesolimbic dopamine system: The final common pathway for the reinforcing effect of drugs of abuse?" *Neuroscience and Biobehavioral Reviews* 30: 215–38

Pinker, S. 2002. *The Blank Slate: The Modern Denial of Human Nature.* New York: Viking Penguin.

Poldrack, R. A., and M. G. Packard. 2003. "Competition between memory systems: converging evidence from animal and human studies." *Neuropsychologia* 41: 245–51.

Prather, M. D., P. Lavenex, M. L. Mauldin-Jourdain, et al. 2001. "Increased social fear and decreased fear of objects in monkeys with neonatal amygdala lesions." *Neuroscience* 106 (4): 653–58.

Raine, A. 1993. *The Psychopathology of Crime: Criminal Behavior as a Clinical Disorder.* London: Academic Press.

Ramachandran, V. S. 1988. "Perception of shape from shading." *Nature* 331 (6152): 163–66.

———.1997. "Why do gentlemen prefer blondes?" *Medical Hypotheses* 48 (1): 19–20.

Ramachandran, V. S., and P. Cavanagh. 1987. "Motion capture anisotropy." *Vision Research* 27 (1): 97–106.

Rao, R. P. 1999. "An optimal estimation approach to visual perception and learning." *Vision Research* 39 (11): 1963–89.

Rauch, S. L., L. M. Shin, and E. A. Phelps. 2006. "Neurocircuitry models of posttraumatic stress disorder and extinction: human neuroimaging research—past, present, and future." *Biological Psychiatry* 60 (4): 376–82.

Raz, A., T. Shapiro, J. Fan, and M. I. Posner. 2002. "Hypnotic suggestion and the modulation of Stroop interference." *Archives of General Psychiatry* 59 (12): 1155–61.

Reichenbach, H. 1951. *The Rise of Scientific Philosophy*. Berkeley: University of California Press.

Reitman, W., R. Nado, and B. Wilcox. 1978. "Machine perception: What makes it so hard for computers to see?" In *Perception and Cognition: Issues in the Foundations of Psychology*, edited by C. W. Savage, 65–87. Volume IX of Minnesota Studies in the Philosophy of Science. Minneapolis: University of Minnesota Press.

Rensink, R. A., J. K. O'Regan, and J. J. Clark. 1997. "To see or not to see: The need for attention to perceive changes in scenes." *Psychological Science* 8 (5): 368–73.

Report to Governor. Charles J. Whitman Catastrophe, Medical Aspects. September 8, 1966. Austin History Center. http://www.ci.austin.tx.us/library/ahc/whitmat.

Rhawn, J. 2000. *Neuropsychiatry, Neuropsychology, Clinical Neuroscience*. New York: Academic Press.

Ritter, M. 2006. "Brain-scan lie detectors coming in near future." Transcript. Fox News, January 31.

Roberts, S. C., J. Havlicek, and J. Flegr. 2004. "Female facial attractiveness increases during the fertile phase of the menstrual cycle." *Proceedings of the Royal Society of London B*, 271 : S270–72.

Robert, S., N. Gray, J. Smith, M. Morris, and M. MacCulloch. 2004. "Implicit affective associations to violence in psychopathic murderers." *Journal of Forensic Psychiatry &Psychology* 15 (4): 620–41.

Robinson, G. E., C. M. Grozinger, and C. W. Whitfield. 2005. "Sociogenomics: Social life in molecular terms." *National Review of Genetics* 6 (4): 257–70.

Rose, S. 1997. *Lifelines: Biology, Freedom, Determinism*. New York: Oxford University Press.

Rosvold, H. E., A. F. Mirsky, and K. H. Pribram. 1954. "Influence of amygdalectomy on social behavior in monkeys." *Journal of Comparative and Physiological Psychology* 47 (3): 173–78.

Rutter, M. 2005. "Environmentally mediated risks for psychopathology: Research strategies and findings." *Journal of the American Academy of Child and Adolescent Psychiatry* 44: 3–18.

Sapolsky, R. M. 2004. "The frontal cortex and the criminal justice system." *Philosophical Transactions of the Royal Society B* 359 (1451): 1787–96.

Scarpa, A., and A. Raine. 2003. "The psychophysiology of antisocial behavior: Interactions with environmental experiences." In *Biosocial Criminology: Challenging Environmentalism's Supremacy*, edited by A. Walsh and L. Ellis. New York: Nova Science.

Schacter, D.L. 1987. "Implicit memory: History and current status." *Journal of Experimental Psychology: Learning, Memory, and Cognition* 13: 501–18.

Schwartz, J., J. Robert-Ribes, and J. P. Escudier. 1998. "Ten years after Summerfield: A taxonomy of models for audio-visual fusion in speech perception." In *Hearing By Eye II*, edited by R. Campbell, B. Dodd, and D. K. Burnham, 85. East Sussex: Psychology Press.

Scott, S. K., A. W. Young, A. J. Calder, D. J. Hellawell, and J. P. Aggleton, and M. Johnson. 1997. "Impaired auditory recognition of fear and anger following bilateral amygdale lesions." *Nature* 385: 254–57.

Scutt, D., and J. T. Manning. 1996. "Symmetry and ovulation in women." *Human Reproduction* 11: 2477–80.

Selten, J. P., E. Cantor-Graae, and R. S. Kahn. 2007. "Migration and schizophrenia." *Current Opinion in Psychiatry* 20 (2): 111–15.

Shams, L., Y. Kamitani, and S. Shimojo 2000. "Illusions: What you see is what you hear." *Nature* 408 (6814): 788.

Sheets-Johnstone, M. 1998. "Consciousness: a natural history." *Journal of Consciousness Studies* 5 (3): 260–94.

Sherrington, C. 1953. *Man on His Nature*. 2nd ed. New York: Doubleday.

Shipley, T. 1964. "Auditory flutter-driving of visual flicker." *Science* 145: 1328–30.

Simons, D. J. 2000. "Current approaches to change blindness." *Visual Cognition* 7: 1–15.

Simons, D. J., and D. T. Levin. 1998. "Failure to detect changes to people during a real-world interaction." *Psychonomic Bulletin & Review* 5 (4): 644–49.

Singer, W. 2004. "Keiner kann anders, als er ist." *Frankfurter Allgemeine Zeitung*, January 8, 2004. (In German.)

Singh, D. 1993. "Adaptive significance of female physical attractiveness: Role of waist-to-hip ratio." *Journal of Personality and Social Psychology* 65: 293–307.

Singh, D. 1994. "Is thin really beautiful and good? Relationship between waist-to-hip ratio (WHR) and female attractiveness." *Personality and Individual Differences* 16: 123–32.

Snowden, R.J., N. S. Gray, J. Smith, M. Morris, and M. J. MacCulloch. 2004. "Implicit affective associations to violence in psychopathic murderers." *Journal of Forensic Psychiatry and Psychology* 15: 620–41.

Soon, C. S., M. Brass, H. J. Heinze, and J. D. Haynes. 2008. "Unconscious determinants of free decisions in the human brain." *Nature Neuroscience* 11 (5): 543–45.

Stanford, M. S., and E. S. Barratt. 1992. "Impulsivity and the multi-impulsive personality disorder." *Personality and Individual Differences* 13 (7): 831–34.

Stanovich, K. E. 1999. *Who is Rational? Studies of Individual Differences in Reasoning.* Mahweh, NJ: Erlbaum.

Stern, K., and M. K. McClintock. 1998. "Regulation of ovulation by human pheromones." *Nature* 392: 177–79.

Stetson, C., X. Cui, P. R. Montague, and D. M. Eagleman. 2006. "Motor-sensory recalibration leads to an illusory reversal of action and sensation." *Neuron* 51 (5): 651–59.

Stetson, C., M. P. Fiesta, and D. M. Eagleman. 2007. "Does time really slow down during a frightening event?" *PLoS One* 2 (12): e1295.

Stuss, D. T., and D. F. Benson. 1986. *The Frontal Lobes.* New York: Raven Press.

Suomi, J. S. 2004. "How gene–environment interactions shape biobehavioral development: Lessons from studies with rhesus monkeys." *Research in Human Development* 3: 205–22.

———. 2006. "Risk, resilience, and gene x environment interactions in rhesus monkeys." *Annals of the New York Academy of Science* 1094: 52–62.

Symonds, C, and I. MacKenzie. 1957. "Bilateral loss of vision from cerebral infarction." *Brain* 80 (4): 415–55.

Terzian, H., and G. D. Ore. 1955. "Syndrome of Klüver and Bucy: Reproduced in man by bilateral removal of the temporal lobes." *Neurology* 5 (6): 373–80.

Tinbergen, N. 1952. "Derived activities: Their causation, biological significance, origin, and emancipation during evolution." *Quarterly Review of Biology* 27: 1–32.

Tom, G., C. Nelson, T. Srzentic, and R. King. 2007. "Mere exposure and the endowment effect on consumer decision making." *Journal of Psychology* 141 (2): 117–25.

Tong, F., M. Meng, R. Blake. 2006. "Neural bases of binocular rivalry." *Trends in Cognitive Sciences* 10: 502–11.

Tramo, M. J., K. Baynes, R. Fendrich, G. R. Mangun, E. A. Phelps, P. A. Reuter-Lorenz, and M. S. Gazzaniga. 1995. "Hemispheric specialization and interhemispheric integration." In *Epilepsy and the Corpus Callosum*. 2nd edition. New York: Plenum Press.

Tresilian, J. R. 1999. "Visually timed action: Time-out for 'Tau'?" *Trends in Cognitive Sciences* 3: 301–10.

Trimble, M., and A. Freeman. 2006. "An investigation of religiosity and the Gastaut-Geschwind syndrome in patients with temporal lobe epilepsy." *Epilepsy and Behaviour* 9 (3): 407–14.

Tulving, E., D. L. Schacter, and H. A. Stark. 1982. "Priming effects in word-fragment completion are independent on recognition memory." *Learning, Memory, and Cognition* 8: 336–41.

Tversky, A., and E. Shafir. 1992. "Choice under conflict: The dynamics of deferred decision." *Psychological Science* 3: 358–61.

Uexküll, Jakob von. 1909. *Umwelt und Innenwelt der Tiere*. Berlin: J. Springer.
———.1934. "Streifzüge durch die Umwelten von Tieren und Menschen". Translated by Claire H. Schiller as "A Stroll through the worlds of animals and men." In *Instinctive Behavior: The Development of a Modern Concept*, edited by Claire H. Schiller, 5–80. New York: International Universities Press, 1957.

Uher, R., and P. McGuffin. 2007. "The moderation by the serotonin trans- porter gene of environmental adversity in the aetiology of mental illness: Review and methodological analysis." *Molecular Psychiatry* 13 (2): 131–46.

Ullman, S. 1995. "Sequence seeking and counter streams: A computational model for bidirectional information flow in the visual cortex." *Cerebral Cortex* 5 (1): 1–11.

Van den Berghe, P. L., and P. Frost. 1986. "Skin color preference, sexual dimorphism and sexual selection: A case of gene culture coevolution?" *Ethnic and Racial Studies* 9: 87–113.

Varendi, H., and R. H. Porter. 2001. "Breast odour as only maternal stim- ulus elicits crawling towards the odour source." *Acta Paediatrica* 90: 372–75.

Vaughn, D. A., and D. M. Eagleman. 2011. "Faces briefly glimpsed are more attractive," forthcoming.

Wason, P. C. 1971. "Natural and contrived experience in a reasoning problem." *Quarterly Journal of Experimental Psychology* 23: 63–71.

Wason, P. C., and D. Shapiro. 1966. "Reasoning." In *New Horizons in Psychology*, edited by B. M. Foss. Harmondsworth: Penguin.

Waxman, S., and N. Geschwind. 1974. "Hypergraphia in temporal lobe epilepsy." *Neurology* 24: 629–37.

Wegner, D. M. 2002. *The Illusion of Conscious Will.* Cambridge, MA: MIT Press.

Weiger, W. A., and D. M. Bear. 1988. "An approach to the neurology of aggression." *Journal of Psychiatric Research* 22: 85–98.

Weiser, M., N. Werbeloff, T. Vishna, R. Yoffe, G. Lubin, M. Shmushkevitch, and M. Davidson. 2008. "Elaboration on immigration and risk for schizophrenia." *Psychological Medicine* 38 (8): 1113–19.

Weiskrantz, L. 1956. "Behavioral changes associated with ablation of the amygdaloid complex in monkeys." *Journal of Comparative and Physiological Psychology* 49 (4): 381–91.

Weiskrantz, L. 1990. "Outlooks for blindsight: Explicit methodologies for implicit processes." *Proceedings of the Royal Society of London* 239: 247–78.

Weiskrantz, L. 1998. *Blindsight: A Case Study and Implications*. Oxford: Oxford University Press.

Weisstaub, N. V., M. Zhou, A. Lira, et al. 2006. "Cortical 5-HT2A receptor signaling modulates anxiety-like behaviors in mice." *Science* 313 (5786): 536–40.

Welch, R. B., L. D. Duttonhurt, and D. H. Warren. 1986. "Contributions of audition and vision to temporal rate perception." *Perception & Psychophysics* 39: 294–300.

Welch, R. B., and D. H. Warren. 1980. "Immediate perceptual response to intersensory discrepancy." *Psychological Bulletin* 88: 638–67.

Wilson, T. 2002. *Strangers to Ourselves: Discovering the Adaptive Unconscious*. Cambridge, MA: Harvard University Press.

Winston, R. 2003. *Human Instinct: How Our Primeval Impulses Shape Our Modern Lives*. London: Bantam Press.

Wheeler, H. R., and T. D. Cutsforth. 1921. "The number forms of a blind subject." *American Journal of Psychology* 32: 21–25.

Wojnowicz, M. T., M. J. Ferguson, R. Dale, and M. J. Spivey. 2009. "The self-organization of explicit attitudes." *Psychological Science* 20 (11): 1428–35.

Wolpert, D. M., and J. R. Flanagan. 2001. "Motor prediction." *Current Biology* 11: R729–32.

Wolpert, D. M., Z. Ghahramani, and M. I. Jordan. 1995. "An internal model for sensorimotor integration." *Science* 269 (5232): 1880–82.

Yarbus, A. L. 1967. "Eye movements during perception of complex objects." In *Eye Movements and Vision*, edited by L. A. Riggs, 171–96. New York: Plenum Press.

Yu, D. W., and G. H. Shepard. 1998. "Is beauty in the eye of the beholder?" *Nature* 396: 321–22.

Zago, M., B. Gianfranco, V. Maffei, M. Iosa, Y. Ivanenko, and F. Lacquaniti. 2004. "Internal models of target motion: Expected dynamics overrides measured kinematics in timing manual interceptions." *Journal of Neurophysiology* 91: 1620–34.

Zeki, S., and O. Goodenough. 2004. "Law and the brain: Introduction." *Philosophical Transactions of the Royal Society of London B: Biological Sciences* 359 (1451): 1661–65.

Zhengwei, Y., and J. C. Schank. 2006. "Women do not synchronize their menstrual cycles." *Human Nature* 17 (4): 434–47.

Zihl, J., D. von Cramon, and N. Mai. 1983. "Selective disturbance of movement vision after bilateral brain damage." *Brain* 106 (Pt. 2): 313–40.

Zihl, J., D. von Cramon, N. Mai and C. Schmid. 1991. "Disturbance of movement vision after bilateral posterior brain damage: Further evidence and follow-up observations." *Brain* 114 (Pt. 5): 2235–52.

Index

absurd, philosophy of 195
actuarial tests 179
airplane spotters 58
Alberts, Alfred 2
Alcaeus of Mytilene 103
alcohol 101–4, 149, 205
alien hand syndrome 131–2,
    163–4
alliterative alliances 61–2
Alzheimer's disease 128–9, 156
amnesia 58–9
amygdala 126, 153
Annihilation of Man (Paul) 194
anosognosia 135–7
anterior singulate cortex 136
Anti-Defamation League 102
anti-Semitism 101–4, 149
Anton's syndrome 50–1
'apperceptive mass' 14
'appetitions' 13
Aquinas, Saint Thomas 12
Aristotle 35
artificial intelligence 89, 107,
    147–8

assumptions 33–4
attraction 4–5, 75–6, 199
    neural preprogramming 90–5
'auditory driving' 47
Augustine 55
automatism 179
    principle of sufficient 170–1
automatization 71–4
autonomic nervous system 67
awareness *see* consciousness;
    knowledge/awareness gap

babies
    genetics 96–7, 99
    preprogramming 83–4
Babylonian Talmud 103
Bach-y-Rita, Paul 39–40, 44
bacteria 208
beauty 90–5, 199
Bechara, Antoine 66
Beckwith, Allen 160
Beckwith, Dallas 160
behavior, human
    brain damage and 154–7, 201–3

disinhibited 155
drugs and 204–5
understanding 196–7
Bell, Charles 14
Benoit, Chris 176, 206
biases, implicit 59–61
Bigelow, Dr Henry Jacob 202
Billings, Ronald 164
Bingham, Lord 171
biological approach 201–9
Blake, William 7
blameworthiness 151–92
brain damage, behavior changes
and 154–7
Charles Whitman (example) 19,
151–4, 171, 174
criminality, nature of 174–7
development paths 157–60
free will 18, 160–71
impulse control 182–6, 198
mental disorders, conceptual
shifts 172–4
see also legal system
blindness 37–9, 78–9
blind spot 32
blindsight 129–30
change 25–6, 28
color 78–9
denial 50–1
instinct 88
motion 36
sensory substitution 39–44
Bohr, Niels 200
bonding 96–100
Braille 41

brain
damage 154–7, 201–3
function 46–7
hemispheres 123–5, 133–4
structure 1–2
tumors 153–5
BrainPort 41
Breuer, Josef 18
Brown, Thomas Graham 44
Bruno, Giordano 11–12
Bucy, Paul 153
Bush, George W. 65

Camus, Albert 195
Caspi, Avshalom 213, 215
Cattell, James McKeen 15–16
change blindness 25–6, 28
chaos theory 169
Charles Bonnet syndrome 45
Chase, Salmon 108
chick sexing 57–8
Chiu, Pearl 183
choices, free will 18, 160–71,
209
Cho, Seung-Hui 177
Christmas banking clubs 119,
121
citalopram 206
Civil Rights Act (1968) 186
Clarke, Arthur C. 224
Clinton, Bill 117
cocaine 205
cognitive reserve 128–9
Cohen, Jonathan 112, 116
Coleridge, Samuel Taylor 7–8

color blindness 78–9
competition 107
conflicted minds 108
consciousness 4
    change in 16
    controlling role 140–4, 194
    degree of 143–4
    free will 167–8
    Freud and 17–18
    hunches 67, 68
    need-to-know basis 28–30
    summaries 6–7, 8–9, 22, 34
    tasks and 73–4
    unconscious, influence on 69–71
    see also knowledge/awareness gap
consensus building 108
containment 217
Copernicus 10
coprolalia 163
Cosmides, Leda 84, 86, 88
cravings 182–3
Crick, Francis 140, 193
criminality
    legal system and 178–82
    nature of 174–7
culpability see blameworthiness

Damasio, Antonio 67–8
Darwin, Charles 17, 83, 193
deafness 84
decision making 5, 114
    free will choices 18, 160–71, 209
dementia, frontotemporal 155–6
democracy of mind 107–9
Dennett, Daniel 35

Derbyshire, John 103
Descartes, René 56
desire 75, 94
    see also attraction
developmental paths 157–60
Devlin, Dean 103–4
Dickinson, Emily 76
disinhibited behavior 155
distance of interaction 113–14
division of labor model 105–7
DNA 97, 210, 211
d'Olivet, Antoine Fabre 75
domains, overlapping 125–127, 128–129
dopamine system 156–7, 205, 206, 211
Douglas, Justice William O. 19, 135
dreaming 44, 45
drugs
    behavior changes and 204–5
    drug addicts 176
dual-process model 109–11
    rational vs emotional system 111–15, 115–18, 131–3

ears 21
Ebbinghaus, Hermann 56
Edison, Thomas 19
Eiseley, Loren 193
electromagnetic radiation 76–7
emergence 217–19
Emerson, Ralph Waldo 199
emotional vs rational system 111–18, 131–3

energy efficiency 72–3
environment
    developmental paths 157–60
    genetics and 180, 212–13,
        215–16
    umwelt 76–86
epilepsy 207–8
equality 186–8
estrogen 90
ethics 122–3, 181–2
evidence-based sentencing 178–9
Evil Dead 2 (movie) 131
evolutionary factors
    approach 148
    goals 76
    psychology 82, 98–9
expectations 48–50, 53, 54, 141
experience 20–54
    deconstructing 20–2
    time, perception of 51–4
    vision see vision
explicit memory 58, 64
eyes see vision

face–vase illusion 31
Farah, Martha 209
feedback
    loops 46–7
    real-time, imaging 183–4
    sensory 44–5, 52
    trial-and-error 58
fertility 90–1, 93–4
fidelity, genetics and 96–100
Fisher, Helen 99
flash effect 47

flexible intelligence 71, 87, 142–3
fluoxetine 172, 206
Forster, E. M. 59
Foxman, Abraham 102
free will choices 18, 160–71, 209
Freeman, Walter 181
Freud, Sigmund 17–19, 110, 118
frontal lobotomies 181–2
frontotemporal dementia 155–6
Fuess, Elaine 153

Gage, Phineas 201–4
Galileo Galilei 9–12, 193, 196, 217
gaze-reading system 84
Gazzaniga, Michael 133–4
genetics 95, 96–100, 158–9, 208,
        209–24
    environment and 180, 212–13,
        215–16
    predisposition 214–16
    vision and 43
Gibson, Mel 19, 101–4, 149–50
'glimpse effect' 92
Goethe, Johann Wolfgang von 7, 11
Goodwin, Doris Kearns 109
Gore, Al 65–6
greater nervous system 219
Greene, Joshua 112

hallucinations 44, 45–6
Harlow, John Martyn 202
Harris, Eric 177
Hawkins, Jeff 141
Heidegger, Martin 195
Helmholtz, Hermann von 33–4

hemispheres, conflicting 123–5, 133–4
Henry VI, Part 2 (Shakespeare) 114
Herbart, Johann Friedrich 14
hippocampus 126
Hirano, Steve 160
homicidal somnambulism 164–5
hormones 97–8, 206, 209
Horowitz, David 103
Hu Jintao, President 109
Human Genome Project 209–10
human life 193–224
  biological approach 201–9
  genetics *see* genetics
  understanding 193–201
hunches 66–9
Huntington's disease 208–9, 210
Husserl, Edmund 195
Hutton, James 193

ideas 14
illusions
  action/sensation reversal 52–4
  flash effect 47
  illusion-of-truth effect 65–6
  sound 47
  visual 20–1, 23, 30–1, 33, 34–7, 194
imaging methods 173–4
immigrant groups 211
implicit factors
  biases 59–61
  egotism 61–4
  memory 55–7, 58, 64–5
impulse control 182–6, 198

inconsequentiality, human 194–5
inference, unconscious 34
infidelity, genetics and 96–100
'innerer schweinehund' 118
instinct 86–90
  blindness 88
intelligence
  artificial 89, 107, 147–8
  flexible 71, 87, 142–3
intentions 83
interaction, distance of 113–14
internal models 48, 50, 51
Internal Revenue Service (IRS)
  phenomenon 119–20
The Interpretation of Dreams
  (Freud) 18
introspection 21, 30, 35, 37, 61, 89, 199, 224
intuition 21
I-want-it-now deals 116–18

Jacobs, Gerald 43
James, William 83, 87, 88, 106
Jaspers, Karl 195
Jaynes, Julian 124–5
Joan of Arc 207
Jones, John 62
Jung, Carl 8

Kahneman, Daniel 115–16
Kasparov, Garry 72
Kauffman, Stuart 218
Kesey, Ken 181
Kierkegaard, Søren 195
Klebold, Dylan 177

Klüver, Heinrich 153
knowledge/awareness gap 55–74
    automatization 71–4
    hunches 66–9
    implicit biases 59–61
    implicit egotism 61–4
    implicit memory 55–7, 58, 64–5
    priming 64–6
    unconscious, conscious influence
        on 69–71
    unconscious learning 57–9
Koch, Christof 140
Kubla Khan (Coleridge) 8

labor, division of 105–7
LaConte, Stephen 183
Landis, Merkel 118–19, 121
Laplace, Pierre-Simon 217
learning
    flexibility of 71
    unconscious 57–9
LeDoux, Joseph 133–4
legal system
    blameworthiness and 169–70,
        176
    brain-compatible 178–82
    equality and 186–8
    free will and 161–2
    sentencing, modifiability and
        188–92, 197–8
leucotomies 181–2
Li Keqiang 109
Libet, Benjamin 167
Lichtenberg, Georg C. 160
Liebniz, Gottfried Wilhelm 13

limbic system 110
Lincoln, Abraham 108–9
lithium 172
lobotomies 181–2
logic 85–6

McBeath, Mike 37
McClure, Sam 116
McGurk effect 48
Mach, Ernst 20–1
Mach bands 21
machine metaphor 15, 16, 17
MacKay, Donald 48–9
MacLean, Paul 110
major histocompatibility complex
    (MHC) 95–6
Mariotte, Edme 31–2
Marlatt, G. Alan 103
Mascall, E. L. 195
materialist viewpoint 204, 223–4
Maxwell, James Clerk 7
May, Mike 37–9, 42
memory 126
    explicit 58
    implicit 55–7, 58, 64–5
    procedural 56
mental disorders, conceptual shifts
    172–4
mental subagents 105
Meyers, Ronald 123
mind-reading system 84
Minsky, Marvin 86, 105–6, 107
modifiability, sentencing and
    188–92, 197–8
Moniz, Egas 181

monogamy 97–8
Montague, Read 72, 138, 199, 205
Montaigne, Michel de 149, 198
motion 126–7
    aftereffect illusion 35
    blindness 36
motor input 53
Mugabe, President Robert 109
Müller, Johannes Peter 14–15
multistable cube 30–1
Le mythe de Sisyphe (Camus) 195

narcotics see drugs
Nathans, Jeremy 43
natural selection 196
neocortex 110
nerves 14
nervous system, greater 219
neurobiology 200
neuroimaging 173–4
neurological disorders 172
neurons 1–2, 4, 34–5
    activity 21–2
    specialization 86–7, 89
    see also preprogramming, neural
neuroplasticity 188–9
neuroscience 19, 189, 192, 193, 223
neurotransmitter system 206, 209
New Essays on Human
    Understanding (Liebniz) 13
Newton, Isaac 217
number forms 81
observation 21, 220
Occam's razor 222

odor, influence of 95–6
One Flew over the Cuckoo's Nest
    (Kesey) 181
opium 7–8
orbitofrontal cortex 154
organic disorders 172
Orgel, Leslie 148
The Origin of Species (Darwin)
    17, 83
overlapping domains (in the brain)
    125–9

parasomnias 165
Parkinson's disease 156–157
Parks, Kenneth 164–166, 169,
    171, 174
paroxetine 206
Pascal, Blaise 76
Paul, Leslie 194
Pelham, Brett 63
Pennebaker, James 145
perception 13
    conscious 48
    'petite perceptions' 13
    of time 51–4
    unanchored 45–6
pheronomes 96
phrenology 130
physical state, of body 67–8
Pink Floyd 8
Pliny the Elder 103
Pope, Alexander 198
practical reasoners 162
pramipexole 156
predictions 49–50

predisposition, genetic 214–16
prefrontal cortex 182–6
preprogramming, neural 83–6
  beauty and 90–5
  see also genetics
priming 64–6
The Principles of Psychology
  (James) 83
prisons 180–1
procedural memory 56
progress, human 196
prosopagnosia 68
pseudohallucinations 45
psychiatric disorders 172
psychiatry 17
psychoanalysis 18
psychogenic disorders 163, 166
psychology, evolutionary 82, 98–9
psychophysics 14
Ptolemy 10

quantum mechanics 193, 196,
  220–1
quantum physics 168, 200

racism 101–4, 149
radio theory 221–2
Raine, Adrian 174, 215
Ramachandran, Vilayanur 91
rational vs emotional system
  109–18, 131–3
realities, different 79–82
real-time feedback, imaging 183–4
reason 87
  practical reasoners 162

rational vs emotional system
  109–18, 131–3
recidivism 178–80
reductionism 204, 216–17,
  219–20, 223
rehabilitation 180–1, 183, 185,
  188, 197
Reichenbach, Hans 220
Repin, Ilya 28–30
reptilian brain 110
retrospective fabrication of stories
  133–40
risperidone 172
Robinson, Eugene 102–3
robots (artificial intelligence) 89,
  107, 147–8
Ryan, Nolan 8–9

Sapolsky, Robert 172, 208
Sarma, Karthik 136
Scarpa, Angela 174, 215
schizophrenia 211
secrets 144–6
self-knowledge 199, 200
self-reflection 183, 185
sensory factors
  blendings (synesthesia) 79–82
  feedback 44–5, 52
  input 49, 53
  substitution 39–44
sentencing
  evidence-based 178–9
  modifiability and 188–92, 197–8
serotonin 206, 212
sertraline 206

Seward, William 108
Shermer, Michael 139
Sherrington, Sir Charles 22
Shestov, Lev 195
Shiller, Robert 117
Siderius Nuncius (Galileo) 10
Singer, Wolf 177
sleepwalking 164–5, 188
smell 95–6, 97
social factors
    attachment 96–100
    interaction 84–6
    policy 196, 198
The Society of Mind (Minsky) 106
society-of-mind 104–7, 148–9
somnambulism 188
    homicidal 164–5
soul 203
specialization 105–6, 148
    neural 86–7, 89
speed 71–2
Sperry, Roger 123
split-brain patients 123–4, 133–4,
    163–4, 166
steroid rage 176
stories, retrospective fabrication
    133–40
Stroop interference 132–3
Suomi, Stephen 212
synesthesia 79–82
    spatial sequence 81

Tacitus 103
team of rivals framework 101–50
    artificial intelligence 89, 147–8

brain hemispheres 123–5, 133–4
consciousness, role of 140–4
democracy of mind 107–9
dual-process model 109–11
Mel Gibson (example) 101–4
overlapping domains 125–7,
    128–9
rational vs emotional system
    111–18, 131–3
retrospective fabrication of
    stories 133–40
robustness 128–30
secrets 144–6
society-of-mind 104–7,
    148–9
Ulysses contracts 118–23, 197
temptations 117–18, 182–3, 197
testosterone 90, 206
thoughts 3–4, 75–100
    beauty 90–5
    conscious/unconscious boundary
        14
    genetics 96–100
    instinct 86–90
    thinking time 15–16
    umwelt (environment) 76–86
    unconscious 18, 194
time
    duration 52
    perception of 51–4
    rational vs emotional systems
        115–18
    Ulysses contracts 118–23, 197
*The Time Taken Up by Cerebral
    Operations* (Cattell) 15

Tooby, John 84, 86, 88
Tourette's syndrome 163
Tranel, Daniel 68
trial-and-error feedback 58
trolley dilemma 111–13
The Truman Show (movie) 77–78,
    99–100
truth 103, 149
    illusion-of-truth effect 65–6
Tsvangirai, Morgan 109
tumors, brain 153–5
Tversky, Amos 115–16
The Twilight Zone 113

Uexküll, Jacob von 77
Ulysses contracts 118–23, 197
umgebung 77, 82
umwelt (environment) 76–86
unconscious 12–16
    Freud and 18–19
    inference 34
    influence of conscious 69–71
    learning 57–9
    thoughts 18, 194
An Unexpected Visitor (Ilya
    Repin) 28–30
vasopressin 97–8, 206
vigilantism 186
virtue 197
viruses 208, 209

vision 21, 22–34
    boundaries of 24
    depth of 24, 33
    illusions 20–1, 23, 30–1, 33,
        34–7, 194
    internal activity 44–51
    learning to see 37–9
    sensory substitution 39–44
visual cortex 34, 48–9

waterfall illusion 35
Watson, James 193
Weber, Ernst Heinrich 14–15
Weihenmayer, Eric 41, 43
Whitman, Charles 19, 151–4, 171,
    174
Whitman, Walt 101, 104, 107
Williams, Edward H. 202
Williams syndrome 84

Xi Jinping 109

Y chromosome 159, 212
Yarbus, Alfred 28
Yarvitz, Mike 103
Your Brain Is (Almost) Perfect
    (Montague) 72

zombie systems 132, 137, 140–2,
    144, 153, 163, 184